Fuels and Petrochemicals Division 2016

Core Programming Area at the 2016 AIChE Annual Meeting

San Francisco, California, USA
13 - 18 November 2016

ISBN: 978-1-5108-3438-5

Printed from e-media with permission by:

Curran Associates, Inc.
57 Morehouse Lane
Red Hook, NY 12571

Printed by Curran Associates, Inc. (2017)

For permission requests, please contact AIChE
at the address below.

AIChE
120 Wall Street, FL 23
New York, NY 10005-4020

Phone: (800) 242-4363
Fax: (203) 775-5177

www.aiche.org

Additional copies of this publication are available from:

Curran Associates, Inc.
57 Morehouse Lane
Red Hook, NY 12571 USA
Phone: 845-758-0400
Fax: 845-758-2633
Email: curran@proceedings.com
Web: www.proceedings.com

TABLE OF CONTENTS

465268 Scheduling and Purchasing Optimization for Olefin Plants with Downstream Constraints

Abstract:

Olefin cracking takes an important role in chemical industry. Efforts are made to develop new technology to increase its production. Multiple cracking furnaces are employed in parallel to convert various hydrocarbon feed stocks to smaller hydrocarbon molecules such as ethylene, propylene, and butadiene. Each furnace is required to be cleaned up periodically because of coke formation in the reaction coils which decays furnace performance. Thus, production scheduling for a furnace system is a complex task, which provides information of feed type, feed flowrate starting and ending time for each batch of each furnace. Due to the large production volume, olefin plants can make hundreds of thousands of profits by optimal scheduling without any investment. The optimal production scheduling can further give direction for raw material purchasing. Currently, most scheduling models focus on maximizing profit without downstream consideration. However, when an ethylene plant is set up, the downstream component separation ability is limited within a certain range. Thus a scheduling and purchasing model should consider not only maximizing its profit but also downstream restrictions. In this paper, a MILP (mixed integer linear programming) model was built for scheduling and purchasing optimization of an olefin plant with consideration of downstream constraints and other constraints, such as non-simultaneous cleaning, recycling ethane relaying cracking. The efficacy of the model is demonstrated by case studies.

466585 a Study of Paraffin Product Distribution in Fischer-Tropsch Synthesis

Abstract:

Natural resources (coal, natural gas, biomass and wastes) can be converted into liquid hydrocarbon fuel via the Fischer–Tropsch Synthesis (FTS) reaction. Although the products of FTS cover an extremely wide range, and include olefins, paraffins and oxygenates with carbon numbers extending from C_1 to more than C_{100}, the major FT products are paraffins and olefins. The Anderson-Schulz-Flory (ASF) distribution model has been used to describe FTS product distribution, where the mole fraction of hydrocarbons is governed by a chain growth probability factor α (Note that in this model $0< \alpha<1$).

Some researchers have chosen to study the ASF distribution of olefins and paraffins separately. The published results of the ASF distribution of paraffin products show deviations from the ideal ASF distribution, such as a high yield of methane; a distinct change in the slope of the line between light hydrocarbons and long chain hydrocarbons; and in some cases, the values of $\alpha >1$ have been observed for light hydrocarbons (C_3-C_5). Unfortunately, none of the kinetic and mechanistic chain growth models can fully explain the paraffin product distribution that have been obtained experimentally. Nowadays, a few researchers try to link the FT products distribution with thermodynamics.

In the current work, a model based on thermodynamic quasi-equilibrium was introduced to describe the paraffin product distribution in FTS. The model was compared to experimental data obtained from two groups of experiments: one using wide ranges of $H_2/CO/CO_2$ mixtures over both cobalt- and iron-based catalysts and the other by co-feeding olefin and paraffin products into the FT reactor with or without feeding CO/H_2. In addition, experimental data from other researchers, covering a wide range of catalysts,

operating conditions and reactor types, was also used to test the model. It is shown that the measured data follows the equilibrium curves quite closely. Thereafter, Aspen simulations, incorporating the thermodynamic quasi-equilibrium model, have been used to simulate the production of FTS by varying the reaction conditions (temperature, pressure, feeding different kinds of olefin and paraffin products into the reactor). The comparison between the Aspen simulation results and the data obtained experimentally revealed that most of the experimental results were quite close to those arrived at through Aspen thermodynamic simulations, which indicate that the paraffin product distribution can be described by quasi-thermodynamic equilibrium.

467770 Al-based Coating for Coke Reduction during Ethane Steam Cracking

Abstract:

Surface engineering to suppress coke formation in cracking furnaces is one of the key research topics in the ethylene industry. Alumina containing coatings have been claimed as being a major step to improve furnace run lengths. Therefore, in this work a commercial coating CoatAlloy^{TM} has been compared to the uncoated Fe-Ni-Cr base alloy. First, the effect of the coating pretreatment was evaluated in an electrobalance set-up during ethane steam cracking under industrially relevant conditions ($T_{gasphase}$ = 1173 K, P_{tot} = 0.1 MPa, X_{C2H6} = 70 %, dilution Î´= 0.33 kg_{H2O}/kg_{HC}). Addition of nitrogen proved to be detrimental for the anti-coking performance of the coating. Additionally, the effect of presulfiding with 500 ppmw DMDS/H_2O, continuous addition of 41 ppmw S/HC of DMDS, and combinations thereof were evaluated. The tested samples were examined using online thermogravimetry, scanning electron microscopy and energy diffractive X-ray for surface and cross-section analysis together with X-ray photoelectron spectroscopy and wavelength-dispersive X-ray spectroscopy for surface analysis. The passivating coating performed substantially better than the reference Fe-Ni-Cr alloy while no effect on the product distribution was observed.

470801 Techno-Economic and Process Safety Analysis of Ethane to Ethylene Production Process through Oxidative Dehydrogenation (ODH) Route

Abstract:

Ethylene is one of the most important building blocks in the chemical industry and provides the starting point for many essential consumer products; e.g. paints, plastics, and pharmaceuticals. Conventionally, ethylene is produced from ethane through steam cracking at the refineries. However, harsh reaction condition (850°C reaction temperature) and potential hazardous by-product H_2 formation through steam cracking cause the need to look for alternative safer routes of ethylene production. Idaho National Laboratory (INL) is working on a novel route (ODH) of ethylene production, which produces water instead of hydrogen as a by-product during ethylene production. Our initial techno-economic analysis indicates that the minimum selling price of ethylene has to be around $1.85/kg to make the ODH process to be profitable. This presentation will describe the effects of the presence of steam (as a diluent) on the ODH process economics, and process energy efficiency. Additionally, a detailed inherent process safety analysis result of the ODH process will be presented to indicate the most crucial units of the process.

461302 Synthesis and Adsorption Performance of Zeolites for Separating Alkanes in Naphtha

Abstract:

The utilization efficiency of naphtha can be improved remarkably by separating group compositions in naphtha based on molecular scale management. In order to meeting the raw material requirements of ethylene industry and aromatics industry, zeolites was prepared and used for the adsorption separation of alkanes including n-paraffins and monomethyl iso-paraffins in naphtha, which is a more reasonable strategy than the separation of n-paraffins from naphtha.

MFI zeolites were synthesized for its adsorption performance of monomethyl iso-paraffins based on theoretical analysis and experimental investigation. The adsorption capacity of the synthesized MFI zeolites for 2-methylpentane was 0.0383 g/g at 25 . The optimal conditions for the synthesis of MFI zeolites were: n(SiO2):n(Al2O3): n(Na2O):n(H2O) = 80:1:7:2450, aging for 36 h at 25 , crystallization for 48 h at 180 and using n-butylamine as a template. The characterization results showed the synthesized MFI zeolites had a high degree of crystallinity. The average pore size of synthesized MFI zeolites was about 0.58 nm.

The adsorption kinetics of monomethyl iso-paraffins on MFI zeolites was investigated. The diffusion coefficients were gained and the diffusion activation energies of 2-methylpentane, 3-methylpentane, 2-methylheptane and 3-methylheptane were 15.57 kJ/mol, 20.49 kJ/mol, 17.82 kJ/mol and 21.68 kJ/mol. The activation energy of long chain isoparaffin was higher than that of short chain isoparaffin and activation energy of 3-methyl-isoparaffins was higher than that of the corresponding 2-methyl-isoparaffins.

The adsorption separation performance of synthesized MFI zeolites was studied in a fixed bed adsorption column. The adsorption capacities of n-paraffins, monomethyl iso-paraffins and aromatics in MFI zeolites were 0.0529 g/g, 0.0158 g/g and 0.023 g/g. The multimethyl iso-paraffins and cycloalkane were not adsorbed. N-paraffins and monomethyl iso-paraffins, which were 30.4% and 11.8% of naphtha were separated by MFI zeolites. The potential aromatic content of raffinate oil increased to 31.63 %, which was 10 percent higher than that of naphtha.

472984 Catalytic Bitumen Partial Upgrading Under Methane Environment

Abstract:

Heavy oil and bitumen extracted from oil sands such as the ones abundant in Canada cannot be directly sent to existing refineries mainly due to its high viscosity. Traditionally, hydrotreating is widely employed to break down the long carbon chain and complicated polycyclic molecular structure of heavy crude to lower the viscosity, and remove the impurities such as sulfur, nitrogen and metals simultaneously. However, such treatment consumes expensive hydrogen and requires extremely high operating pressure. Replacing hydrogen with methane, the principal component of natural gas, and lowering the operating pressure would significantly decrease the cost of oil refining. The activation and conversion of methane has been a great challenge due to its symmetric molecular structure and high C-H bond energy. The presence of higher hydrocarbons, which is rich during the upgrading process of heavy oil, would significantly enhance the conversion of methane. In the present work, methane is activated with ZSM-5 based catalysts loaded with active metals and upgrade the heavy oil at low pressure. The charged catalyst witnesses significant viscosity reduction, excellent product stability and compatibility, decreased total acid number, averaged molecular weight, asphaltene content, and increased H/C atomic ratio of the product. The reaction between the heavy oil molecules and methane is studied by choosing several representative model compounds including n-butylbenzene to analog the feedstock components in heavy oil. The study of the reaction is executed from multiple aspects with several analytic instruments, including the GC-MS analysis of products, which provides the composition information, and Diffuse Reflectance Infrared Fourier Transform (DRIFT) Spectroscopy, which detects the formation of organic species on the surface of the catalyst during the upgrading process. ^{13}C and ^{2}D enriched methane are also employed to track the catalytic pathways. Several key products such as ethylbenzene, propylbenzene, isopropylbenzene and butylbenzene, which are the saturation products of styrene, are separated and purified. Their ^{1}H NMR, ^{13}C NMR, FTIR and GC-MS spectra reveal how the CH_x and H are added into unsaturated bonds. This work clearly demonstrates the feasibility of upgrading heavy oil by directly using cheap natural gas on zeolite-supported catalyst at moderate pressure. This mechanistic study would also benefit the rational design of catalyst for direct upgrading of heavy crude oil using natural gas in the future.

472197 Molecular Characterziation and Modeling for Feedstocks Including Heavy Oil and Reactors

Abstract:

Simulation of refinery-wide processes helps the petroleum industry to design, optimize and operate the refineries profitably and safely, while meeting all necessary regulatory requirements. The traditional pseudo-component approach developed over 30 years ago is still widely used today to represent the feedstocks and products in these process models. This approach uses correlations and heuristics to provide thermodynamic and other properties required in the simulation. For refinery reactor models, kinetic lumps are used to represent molecular classes that undergo reactions to produce the desired products. These kinetic lumps are determined from either finger-print data of the feed or correlations that are based on other properties (such as PNA or sulfur contents) of the feed. Both the pseudo-component and kinetic lump representations do not contain molecular level information. Due to the increasing demands of environment regulations, changing feedstocks, higher margins, etc., the petroleum industry is very interested in improving the accuracy and understanding of the refinery process models and has been exploring the use molecular information for quite some time. In this presentation, we will describe the molecular approach recently developed to leverage molecular knowledge to increase the accuracy of process simulation. The molecular method was applied to characterize crude oils using traditional laboratory assay data (quantitatively match) as well as analytical data such as Fourier Transform Ion Cyclotron Resonance Mass Spectrometry (FT-ICR-MS) (qualitatively match). The resulting molecular compounds, compositions and associated properties are then used to calculate the properties of the process streams. The molecular information are also used directly to represent kinetic lumps composition used in refinery reactor models.

471382 An Innovative Technique for Oxy-Cracking of Quinolin-65 As a Model Heavy Hydrocarbon Compound

Abstract:

Oil refining processes typically involves a coking process that results in a carbon rich solid waste hydrocarbon byproduct called petroleum coke (petcoke). The production rate of the petcoke is over 20 million tonnes per day. Recently, petcoke from Canadian oil sands in U.S. refineries has led to the appearance of large stockpiles adjacent to urban communities in Detroit and Chicago, because limited markets exist for this commodity. The petcoke has strong effects on human and animal health. The main threat to urban populations in the vicinity of petcoke piles is most likely fugitive dust emissions in the form of fine particulate matter. Current industrial processes for treating these kinds of such complicated waste hydrocarbons like direct gasification and combustion have environmental drawbacks centralized mainly in the high production of gaseous CO_2.

In this study, we propose a nano-catalyzed oxy-cracking process that works at mild operating temperature and pressure for high solubilisation of petcoke in relatively low proportions of water. This would open the possibilities for an efficient disposition of petcoke which could be converted at low processing costs into a transportable stream or, through their selective adsorption, be anchored on active catalysts for converting them into lighter products, or alternatively be used efficiently as hydrogen generating source via catalytic steam gasification at much lower temperature than the conventional gasification process.

To understand the mechanism and the kinetics of such innovative technology, Quinolin-65 (Q-65) was selected as a model-oxy-cracking compound mimicking the solid waste hydrocarbons in aqueous-based solutions. In the first phase of this study, the reaction took place in a non-catalyzed batch process operated at optimized oxygen partial pressure and temperature. The preliminary results enabled us to propose a reaction pathway based on the radical reaction mechanism to describe the kinetic behavior of the oxy-cracked Q-65. Hence, kinetic experimental results indicated that

the Q-65 oxidation undergoes a parallel consecutive reaction in which an oxidative decomposition took place in the first step producing different aromatic intermediates as confirmed by the NMR analysis. Further, the intermediates were oxy-cracked consecutively into different families of organic acids and small amount of solubilized CO_2 as confirmed by the inorganic carbons analysis. To support our experimental findings, a theoretical computational modeling is conducted. Therefore, the complete radical reaction mechanism will be discussed and elaborated.

465284 Upgrading of Petroleum Vacuum Residue Using Supercritical Hydrocarbon Solvents in Batch and Continuous Systems

Abstract:

As demands for transport fuels and lighter oils have increased, petroleum industries have been forced to refine low-quality extra heavy oils. Among the heavy feedstocks, petroleum vacuum residue (VR), which is the heaviest fraction of crude oil, has high boiling points over 520 °C and contains about 5 wt. % of sulfur, 0.5 wt. % of nitrogen, and organometallic compounds such as nickel and vanadium. Generally, heavy feedstocks can be converted into light products by either carbon rejection process or hydro-conversion process. However, it is difficult for extra heavy oils to be converted into liquid fuels by the conventional upgrading processes due to the high concentration of impurities and asphaltenes.

In the study, supercritical upgrading process (SUP) was applied to upgrade vacuum residue (VR) to light products with small amount of coke formation in batch and continuous systems. Since the physical and transport properties at near-critical region of fluid are altered sensitively, the parametric study for the SUP was required to elucidate optimum operating conditions, depending on various hydrocarbon solvents. The upgrading results were evaluated on residue conversion, coke formation, and product distribution (naphtha, middle distillate, vacuum gas oil, and residue) at various operating conditions; temperature (350-450 °C), hydrogen partial pressure (0-3.45 MPa), reaction time (5-30 min). In addition, further study on a continuous system was conducted on the basis of the results from the batch system. In the continuous system, residence time worked as one of key operating parameters to suppress the coke formation.

470536 The Binary Pyrolysis of Sulfur Compounds with Alkyl Aromatics and n-Alkanes

Abstract:

Petroleum refiners need reliable kinetic models for the optimization of oil residue conversion processes. Modeling of chemical reactions involving heavy ends of petroleum is complicated. Structures and mixtures are very complex, and dissimilar depending on the origin of crudes.

Countless species differ in reactivity should be accounted for in these models, which in turn needs explicit molecular detail of the feed. During conversion many structural changes take place, high boiling compounds broken down into smaller fragments; all sulfur as well as metals such as vanadium and nickel are removed and the most of the aromatics turn into naphthenes which further turn into paraffins and iso-paraffins.

Pure hydrocarbon model compounds pyrolysis gives important information about the reaction kinetics. The purpose of model compounds pyrolysis and its components is to reveal the actual chemical kinetics of the molecules that play a fundamental role during the conversion process. Binary pyrolysis of pure hydrocarbons obtained from the reaction of sulfur compounds reactions will provide important information about heavy oil conversion reactions.

In this study, the binary pyrolysis of decilmercaptane, thiophene n-tetradecane and butyl benzene was investigated at reaction temperatures of of 400-500 °C temperature for reaction time ranging from 30 min to 180 min. In the second phase with the same reaction conditions the binary compounds pyrolysis was carried out. Reaction rate constants and arrhenius parameters for the pyrolysis of binary mixtures obtained from the reaction model were compared with pure compounds.

468016 Use of Unconventional Resources in State of the Art Oil Refining Processes

Abstract:

Conventional oils which are readily accessible and homogenous compared to the unconventional oils are running out. This scarcity draws growing attention to unconventional oils in order to meet the energy requirement of the world. On the other hand, unconventional oils are heavy and complex structures which are tightly trapped between or bound to sand or rock; therefore, new challenges emerge to convert these oils into more valuable products.

In this study, thermal upgrading of oil sand and heavy bitumen derived from domestic reserves is studied to reveal the processibility of these feedstocks in delayed coker units. In addition to that, these feedstocks are blended with highly aromatic refinery by-products to evaluate the effect of aromaticity on coke and liquid yield as well as coke morphology. Feedstock of the delayed coker unit from a refinery is used as a reference for both upgrading and characterization parts.

Oil sands are subjected to solvent extraction method with toluene to separate sand and hydrocarbon samples. After that, these unconventional feedstocks are evaluated for their elemental composition, SARA content, number and weight average molecular weight, specific gravity, dynamic viscosity and S-value (an indicator for stability of asphaltene molecules and coke morphology). Thermal upgrading reactions are conducted at 500 °C for 90 minutes in a tubing bomb reactor having a 10 mL internal volume. Both neat feedstocks and the ones blended with refinery by-products are used in reactions. Coke and liquid products are characterized by means of scanning electron microscopy (SEM) and gas chromatography (GC-SIMDIS), respectively. Moreover, liquid products are further analyzed for their elemental composition, SARA content, molecular weight, specific gravity and dynamic viscosity.

The characterization results indicated that oil sands and heavy bitumen can be used in delayed coker reactor systems but they should be blended with conventional oils to meet design requirements of the units in terms of sulfur and asphaltene content or specific gravity. Furthermore, thermal upgrading reactions demonstrated that unconventional and reference feedstock give nearly same coke yield (~20% wt.) but their liquid product distributions are quite different from each other. Addition of refinery by-products provides around 10% decrease in coke yield which can be attributed to the prevention of phase separation during thermal upgrading reaction by aromatic compounds which can dissolve maltene fractions and asphaltene precursors. Thereby, polycondensation reactions are retarded and coke yield is decreased. Moreover, the S-value results and microstructures from SEM analysis indicated that addition of highly aromatic refinery by-products affects the coke morphology, which specifies the value of coke and occurrence of hot-spot problems in delayed coker units.

Knowing the fact that liquid products are more valuable as compared to solid coke samples whose annual production capacities may reach up to million tons/year for a single refinery, decreasing coke yield and changing coke morphology with easy-to-implement solutions is a powerful tool to increase the profit of oil refineries in a competitive global market.

449940 Developing a Dead Heavy Oil Viscosity Correlation for Lower Fars, Kuwait

Abstract:

Abstract

Viscosity of the produced fluid is an important property for oil and gas industry. It is used in calculations of fluid flow in reservoirs, designing production tubing and flow lines. Accurate estimation of heavy oil viscosity is the most important, as it is one of the key parameters in evaluating a given prospect for the possibility of cold oil production and planning the type of thermal Enhanced Oil Recovery (EOR) methods such as hot water, steam cycling, steam flooding, steam assisted gravity drainage or in-situ combustion. Viscosity of crude oil is usually measured at reservoir temperature at various pressures using bottom-hole sample or surface recombined sample.

Occasionally, the sample is difficult to obtain or becomes unreliable. In this case calculation methods are used to estimate the crude oil viscosity as a function of easily measured production information such as API gravity or detailed compositional data. Many correlations have been presented to estimate crude oil viscosity for crude oils of a certain field or region. Application of these correlations to other parts or regions results in large error in estimated viscosity because each crude oil has its own physico-chemical characteristics that makes it different from other crudes.

In this study, the accuracy of the well-known correlations are tested against heavy oil viscosity data collected from Lower Fars formation in Kuwait. Later, a new correlation is presented to estimate the dead oil viscosity of Lower Fars. Accuracy and reliability of the proposed model is compared to published correlations such as Beal (1946), GlasØ (1980), Kartoatmodjo and Schmidt (1994), and Al-Omair et al. (2014).The results showed that the new correlation is able to predict the dead oil viscosity with mean absolute relative error of 0.16, correlation coefficient equal of 96%, and standard deviation of 0.17.

473688 New World of Opportunity Crudes

Abstract:

The refining industry has changed noticeably over the past several years with an ample supply of opportunity crude oils available, which include light tight oils and heavy sour crudes, and sometimes high TAN crudes. Opportunity crudes have been around for years, but only until recently has the abundance of these discounted crudes shifted the behavior of refiners.

With the use of opportunity crude oils comes new processing challenges; Automation technology is advancing to meet these challenges to ensure refiners are successfully alleviating these new issues. Advances in real time crude analysis and crude unit monitoring and control allow refiners to maximize opportunity crude processing while mitigating issues like accelerated fouling and corrosion.

Refiners on the coast will typically have access to more opportunity crudes than those landlocked. Because crude oils can be very different in properties, refiners try to match the required crude oil composition for their refinery configuration; usually by blending two or more crude oils. To make matters more challenging, sometimes crude oils purchased to process are crudes that the refiner has no experience processing.

Crude blending operation has become more important with the ability to establish consistent crude feed properties. Steady operation enables better optimization of the crude unit by pushing against the appropriate constraints which often includes maximizing the quantity of discounted opportunity crude oils blended.

This presentation will highlight on new operating challenges, and show a couple examples where refiners have made modifications in both technology and behavior to adapt to the new world of opportunity crudes.

463572 Hydrocarbon Mixture Bubble Points in Nanoporous Silica Monoliths

Abstract:

Shale reservoirs are now becoming the focus of new oil and gas exploration and development. These shale reservoirs usually have lower permeability and porosity with a pore size distribution ranging between 1-20 nm. It is hypothesized that fluids in the confined systems (shale reservoirs) have different thermo-dynamic properties than conventional reservoirs. However, because of the difficulty of the realization of the pore structure of shales, there have been few experimental studies. In my research, two different nanoporous silica based monoliths were synthesized with similar pore size ranges as the shale reservoirs, represented in the porous system of shale reservoirs. Monoliths can be synthesized into a desired shape by putting them in a customized container. Because of this advantage, the synthesized monoliths have only nano-sized pores, not macropores. Two nanoporous monoliths were synthesized by modifying the synthesizing procedures that could be used to synthesize hexagonal and cubic mesoporous silica powders, which have different morphologies. From comparing the saturation pressure with and without nanoporous media, we can confirm the hypothesis that nano-sized pores have an effect on saturation pressure. Also, we can see the different morphologies of the two monoliths have an effect on saturation pressure. These synthesized monoliths have been characterized by X-ray diffraction (XRD), nitrogen adsorption/desorption isotherm (BET), and transmission electron microscopy (TEM), which show their physicochemical properties. The phase diagram of methane and decane mixture has been modeled with the Winprop module from the CMG reservoir simulator. The experimentally measured saturation pressure matches with the simulated result. However, the measured saturation pressure changes with nanoporous media.

455711 Forecasting Gas-Oil Ratios and Solution Gas Production from Liquid Rich Shale Reservoirs

Abstract:

The basic issue: To enhance our knowledge about the production mechanisms of shale oil reservoirs and better exploit their economic potential, it is important to be able to forecast production of the secondary phase, namely gas.

Why it is important: Stakeholders often focus on oil forecasts when it comes to shale oil reservoirs, ignoring the equally important solution gas produced from these plays. Lack of production data, complicated flow mechanisms of liquid rich shale reservoirs, production pattern of producing gas-oil ratios among other factors, make the task of forecasting solution gas production difficult.

What we are proposing: The objective of this paper is to present a new method for estimating producing gas-oil ratios (GOR) and solution gas production from shale volatile oil reservoirs. The method is data driven while observing the fundamental equations that govern hydrocarbon production over time. It entails the following steps:

Step 1 – Generate representative collection of data: To develop the functional form that describes the evolution of GOR over time for a producing well, run a large number of representative simulations over a fairly long production time (e.g. 30 years) so that a variety of wells in various reservoirs under diverse producing conditions can be well represented in the sample.

Step 2 – Identify natural GOR shapes through PCA: Stack GOR profiles on top of each other, to construct a matrix of dimension perform principal component analysis (PCA) on , and then approximate as , where . The vectors can be thought of as natural shapes over time, whose linear combination can reproduce – with reasonable approximation – every conceivable GOR profile over time.

Note:Steps 1 and 2 will have to be completed perhaps only once or at worst a limited number of times, since the natural GOR shapes they produce will be universal.

Step 3 – Fit a linear combination of natural GOR shapes to data over limited time: Given GOR data for a limited production history over time , identify estimates by finding .

Step 4 – Use the model of Step 3 for future predictions: The value of GOR at a future time along with a corresponding confidence interval can be estimated using standard formulas as where is the information matrix.

Case study: A commercial compositional simulator was used to simulate 30 years of production from a shale volatile oil reservoir with 4 different reservoir fluids. The reservoir permeability and porosity were 0.001 md and 6% respectively. Several values for the following parameters were considered in a total of the simulations:

1. 1. Critical gas saturations between 5% and 20%;
2. 2. Constant bottomhole pressures (BHP) between 500 and 1500 psi; and
3. 3. Initial reservoir pressures between 3000 and 5000 psi.

PCA on the GOR data produced by these simulations revealed (through cross-validation) that a small number of principal components adequately capture the GOR profiles observed over the 30-year period. To test how well the resulting model can forecast future production from production data over limited time, we considered a horizontal well of length 5000 ft, with 20 hydraulic fractures of 150 ft half-length, spaced 250 ft apart. We simulated producing GOR over 30 years, fit the model to the corresponding simulated data over periods from 6 months to 3 years, and then compared GOR forecasts by the model to GOR values produced by the simulator into the future. The future profiles produced by the PCA model matched those produced by the simulator with reasonable level of accuracy. The PCA model was also found to match the simulator on solution gas production, calculated by integrating producing GOR over cumulative oil production.

Conclusions: This work presents a novel approach of forecasting production from liquid rich shale plays, based on a well established statistical approach. It bypasses the need to invent an ad hoc functional form for GOR over time that would be both simple for computation and realistic enough to capture real data. The proposed approach can help improve reserves estimation, reservoir management, and project economics.

467020 Solvent Extraction for Bitumen Recovery from Oil Sands Ores

Abstract:

Unconventional oils, including oil/tar sands, shale oil, heavy oil, etc., have been considered as an alternative of traditional crude oils due to their huge reserves (about two thirds of the total oils) buried around the world. It is estimated that there are about 1700 trillion barrels of bitumen contained in the oil sands ores in Alberta, Canada(Chew, 2014; Gray et al., 2009). To release this kind of petroleum, hot water-based extraction (HWBE) has been proposed and applied in industry which creates large volumes of wet tailings with fine solids. To avoid this environment pollution, the solvent extraction has been attracting increasing attention from industry due to its high extraction efficiency, less water and energy consumption, and high compatibility for different kinds of ores.

In the present study, solvent extraction of oil sands has been investigated using pure organic solvents, including toluene, tetrahydrofuran (THF), cyclohexane, n-heptane, n-hexane, ethyl acetate, and etc. Results of bottle tests showed that the bitumen recovery increased with the increase of the number of carbon and the cyclic hydrocarbons (toluene, THF and cyclohexane) performed better than alkane hydrocarbons (n-heptane, n-hexane, ethyl acetate and acetone). To analysis the bitumen in the extracted phase more detailedly, it has been divided into four fractions, named as saturates, aromatics, resins, and asphaltenes (SARA). Results show that increasing the aromaticity of the solvent is beneficial for the recovery of heavy fractions (e.g., resins and asphaltenes) of bitumen from the oil sands.

To further explore the solvent extraction process, solubility simulation in COSMOtherm software was applied and four typical molecules were selected as the representatives of SARA. A similar trend of bitumen recovery with different solvents was obtained as an order of: THF â‰¥ toluene > cyclohexane > n-heptane > n-hexane > ethyl acetate. Firstly, the efficiency of solvent extraction is found to be highly dependent on the polarity of the solvent. The solvents with too low (i.e. n-hexane and n-heptane) or too high (i.e. acetone and methanol) polarity had a poor solubility to bitumen. The polarity of the solvent will cause certain effect to the bitumen recovery because some special polar groups (i.e. groups containing oxygen, nitrogen, sulfur, and etc.) affected the properties of bitumen components. Secondly, the enhancement of solvent aromaticity makes the recovery of aromatics, resins and asphaltenes increase. The reason for these may be that the heavy bituminous constituents contain aromatic or cyclic structures. From the solubility point of view, the more similarity of microstructure between bitumen components and solvent, the more bitumen was recovered. Finally, after careful analysis, it is found that the dispersion force of the solvent plays a dominate role (weight > 55%) in dissolving bitumen fractions, followed by hydrogen-bonding force (weight at 0~26%) and polar force (weight at 0~18%). This finding allows us to consider more detailed influence factors when selecting the potential solvents for bitumen recovery process and solubility simulation may be a relatively efficient and accurate previous prediction of solvents screening.

Accordingly, the solvent extraction process would be a promising method for the recovery of unconventional oil from the ores. This study would improve the knowledge base for bitumen recovery mechanism and serve for future work on the solvent extraction technology applications.

References:

Chew, K.J., 2014. The future of oil: unconventional fossil fuels. Philos Trans A Math Phys Eng Sci 372, 20120324.

Gray, M., Xu, Z., Masliyah, J., 2009. Physics in the oil sands of Alberta. Phys. Today 62, 31-35.

464457 Short Residue to Useful Products: Different Pathways

Abstract:

With the depletion of the oil reserves, the oil is getting heavy day by day. The utilization of the heavy oil in the refinery process leads to the generation of high amount of short residue from the vacuum distillation units. The present study is about finding the different pathways for the utilization of the short residue to the formation of useful products. The conversion of short residue to useful product also helps in increasing the economic viability of the refinery processes. This research work initially studies about the quality of the various crudes available in the Indian petroleum industry. The different processes involved for the upgradation are categorized as Carbon Rejection Process and Hydrogen Addition Process. The present work further discussed on the selection of the specific process depending on the quality of the short residue left in the refinery.

466084 Catalyst Performance Evaluation in Catalytic Naphtha Reforming - Using State of the Art Experimental Design and Reactor Technology to Select Catalysts for Non-Isothermal Reactors in Industry

Abstract:

Catalyst Performance Evaluation in Catalytic Naphtha Reforming - Using State of the Art Experimental Design and Reactor Technology to Select Catalysts for Non-Isothermal Reactors in Industry

Erik-Jan Ras, Roel Moonen, Pieter Imhof, Stephan Janbroers, Markus Leuenberger, Tom Huizinga, Rien Elfring and Mark D. Moser

Avantium Chemicals B.V., Zekeringstraat 29, 1014 BV Amsterdam, The Netherlands

Key message

- Catalyst selection decisions can be made with confidence, based on highly accurate and reliable data
- Proven track record in equipment design and test design for the catalytic reforming of naphtha feedstocks
- Our experience in the design of test equipment and protocols is key, applicable to continuous regeneration, semi-regenerative and cyclic mode of operation

Abstract

With ever changing specifications for gasoline, and an increasing demand for aromatics in general, catalytic reforming of naphtha remains of key importance in a modern refinery. This demanding application is characterized by a set of distinct challenges. From a reactor engineering perspective – all technologies currently employed at commercial scale have in common that they are operated in a non-isothermal manner. From a catalyst perspective, each catalyst shows a significant degree of deactivation during the time period needed for a typical performance test. On top of these challenges, the differences between commercial catalysts in terms of performance are typically small.

Using a parallel reactor set up designed specifically to meet these challenges, we can map catalyst performance, and project performance at large scale successfully. Combining aspects like true isothermicity, and an accurate control of space velocity and reactor pressure, the performance data obtained is unprecedented in terms of accuracy. Moreover, within the same reactor setup, we provide different modes of operation – isothermal operation as well as iso-octane operation. These key features, in combination with the use of statistical experimental design methodology provide all the information needed to select the right catalyst based on key performance indicators as well as their confidence intervals.

This presentation will highlight the intimate relationship between the design of a test, and the equipment used for a test. Moreover, the workflow required for data processing – transforming raw data to information – will be discussed. Topics will include key performance indicators (activity, $C5^+$ yield, hydrogen production), but will also show that during the same test a fully detailed hydrocarbon analysis is obtained at the same level of accuracy. This analysis provides additional insight of the value of the reformate produced. The presentation will use examples from real world projects.

FACTORS AFFECTING H_2S SCAVENGER PERFORMANCE IN ASPHALT

Kyle Cattanach, Weldon Cappel, Don Wolfe
Baker Hughes Incorporated

ABSTRACT

Hydrogen Sulfide (H_2S) presents a very serious threat to the health and safety of individuals involved in the transport and storage of asphalt. H_2S scavengers are employed to mitigate the presence of H_2S in asphalt transport and storage conditions were it may be present. H_2S scavenger performance and effectiveness has been shown to vary significantly at different field locations. This study will examine and document the critical factors in asphalt that can lead to varying H_2S scavenger performance in terms of removal of H_2S. Laboratory studies of H_2S scavenger performance were done using a custom built testing apparatus. Industry-representative samples of asphalt were collected from three different locations and then heated to various temperatures in the laboratory to observe and record the H_2S generation profiles in the vapor phase. Neat asphalt samples from various locations were heated, and the H_2S generation profiles were compared. The various metal-based H_2S scavengers tested were not negatively impacted by increasing the asphalt temperature up to 425°F. Laboratory tests indicate that the higher temperatures improved the scavenger performance, presumably due to improved mixing and contact efficiency between the H_2S and the H_2S scavenger.

Keywords: H_2S, asphalt, scavengers, mixing, sulfur, vapor

INTRODUCTION

The health and safety of personal and operators is of paramount importance in removing H_2S from asphalt and other fluid streams[1-4]. This study was initiated to better understand the H_2S removal performance of a service companies H_2S scavenger products and to generate quantitative data in the laboratory under controlled conditions of the phenomena of H_2S generation in sour asphalt. Specifically, a metal-based H_2S scavenger product line was tested. Among the critical findings are: as the temperature of the asphalt sample increases, H_2S is released to the vapor phase by degradation of sulfur compounds in the asphalt sample. However, the performance of the metals based H_2S scavenger is not significantly degraded by the increased temperature. The laboratory findings confirm observations in the field that mixing, contact time and H_2S scavenger concentration are critical to H_2S scavenger performance. These results confirm those of previous studies[1,3,4]. It was observed that the increase in vapor phase H_2S competes against the improved mixing and contact efficiency at the higher temperatures, which leads to improved scavenger performance. Solvent flashing of H_2S scavengers at higher temperatures does not significantly degrade H_2S scavenger performance. The handling properties afforded by the addition of a solvent package to the active component of the H_2S scavenger far outweigh any potential negative impacts on H_2S scavenger efficacy due to solvent loss or flashing. Throughout this testing great care was taken to maximize the data generated with the limited amount of asphalt samples received from the field. In many cases, conclusions were made in the absence of repeated data sets. This was due to the limited amount of sour asphalt available from the field for testing as well as the need to reduce tests due to the inherent safety risks of testing sour asphalt at elevated temperatures.

EXPERIMENTAL

A 1000-ml flask was placed in a heating mantle with a condenser (see Figure 1) for all tests run. The flask is sealed with a PTFE gasket and has a three neck connecter on top. For standard tests 500 ml of the previously heated asphalt samples were then placed in the flask containing a stir bar. The asphalt sample was then heated to the listed test temperatures. The highest temperature tested was 425°F due to safety concerns, as the asphalt began to splatter at higher temps. The asphalt sample was kept at each temperature for one hour prior to measurements. The vapor phase measurements were then either taken by a Drager Tube or gas chromatograph (GC).

Initial experiments were carried out by Drager tube measurements in the headspace of the reaction flask as this is an industry-wide accepted method for measurement of headspace H_2S. An Inficon® 3000 Micro GC with sulfur specific column was used for all GC measurements. A needle was placed inside the vessel directly above the surface of the asphalt to draw up vapor samples. Measurements were periodically spot-checked via Drager tubes and were found to be in agreement from 1-5%.

An internally developed liquid phase H_2S testing methodology was developed. A heated syringe was used to uptake an aliquot of the asphalt sample. The asphalt sample was then placed in a pre sparged sparging flask containing toluene, and the syringe was weighted before and after to determine the precise amount of asphalt added. The asphalt in toluene was then sparged with nitrogen and the H_2S levels were measured via Drager tube. The liquid-level H_2S concentration was then determined based on the mass of asphalt added. Figure 1 shows an image of the testing vessel and liquid H_2S levels set up.

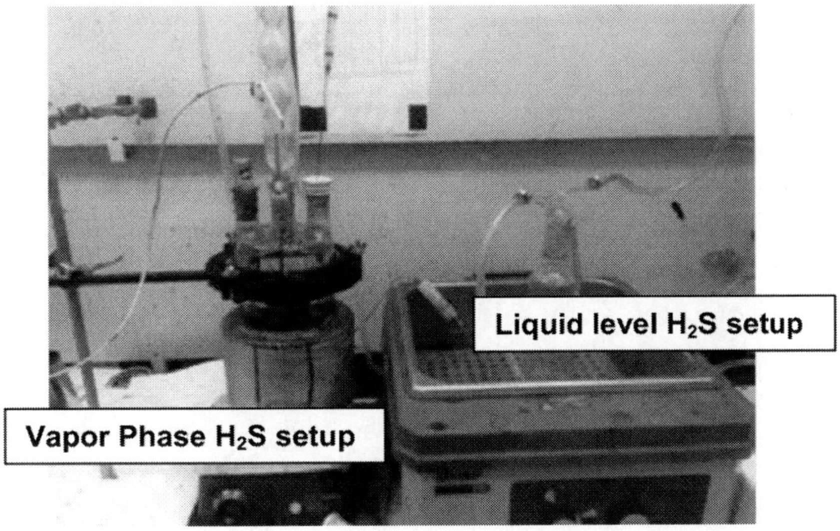

Figure 1: Asphalt test vessel and liquid H_2S level testing set up

RESULTS AND DISSCUSION

H_2S Emission profiles

In order to better understand H_2S scavenger performance, one of the goals of this study was to demonstrate under controlled laboratory conditions generation of H_2S in asphalt via degradation of sulfur compounds. Figure 2 shows H_2S generation profiles from asphalt from three different refinery locations. The emission profiles in Figure 2 demonstrate that as the temperature of the asphalt increases, the amount of H_2S in the vapor phase also increases, presumably to due degradation of sulfur compounds. It stands to reason that as the temperatures of the sample increases, the rate of sulfur compound degradation increases as well and thus the observed increase in vapor phase H_2S.

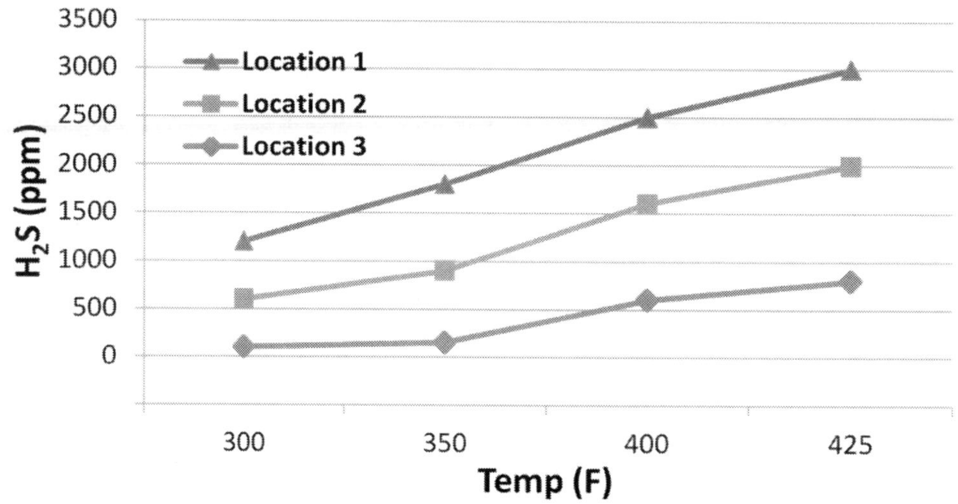

Figure 2: Vapor phase H_2S emission profile for asphalt from three different locations

Figure 2 also reveals the critical factor that asphalt from different refinery locations can have drastically H_2S emissions profiles. While the exact mechanism for the varying H_2S emissions profiles are beyond the scope of this study, it is believed that processing techniques unique to each location play a role in determining the quantity and composition of sulfur species in the asphalt. In addition, thermal degradation studies have shown the tendency of the certain sulfur compounds to degrade more than others in this temperature range[5]. The specific properties of the asphalt and H_2S generation profile of the asphalt at each specific location will have to be taken into account in order to have a successful H_2S scavenger injection program.

H_2S Scavenger testing

Accurate data determining the performance of metal-based H_2S scavengers was collected. The goal was to collect quantitative, reproducible data, in the laboratory and the field, of a metal-based H_2S scavenger product, specifically at elevated temperatures i.e. 350°F and higher. Figure 3 shows the performance of a metals based H_2S scavenger from location 1 at different concentrations. For the tests in Figure 3 asphalt with approximately 3000 ppm of vapor phase H_2S at 425 °F was run neat, and then dosed with 300 ppm (10:1 vapor phase H_2S : H_2S scavenger) and 900 ppm (5:1 vapor phase H_2S : H_2S scavenger) of a metals based scavenger. The dosages were chosen based on typical injection rates of the metal based H_2S scavengers observed in the lab and the field.

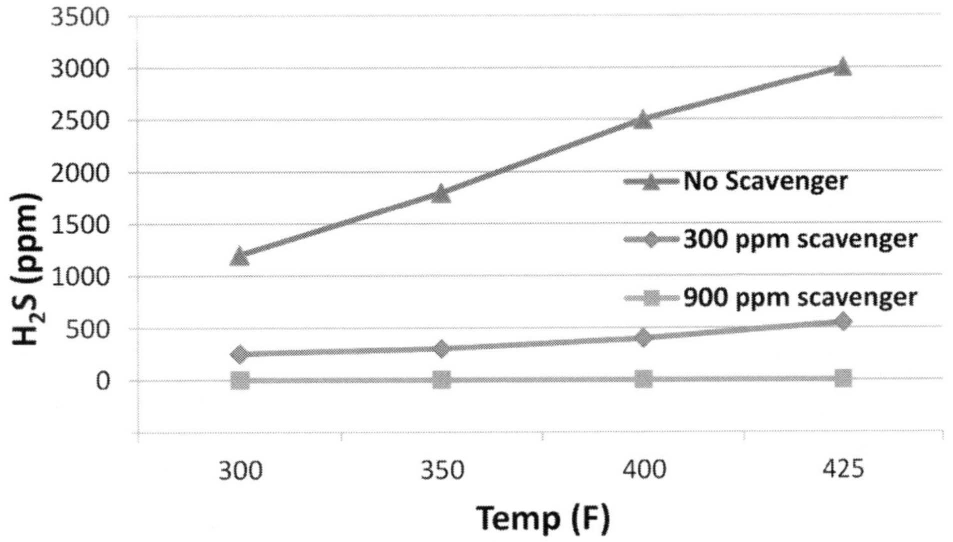

Figure 3: H_2S levels in asphalt from Location 3 with varying concentrations of metal based H_2S scavenger

As Figure 3 demonstrates, as the concentration of the H_2S scavenger increases the vapor phase H_2S levels decreases. With the 900 ppm dosage rate reducing the H_2S level in the vapor phase to near undectable levels (100% effective). As the H_2S scavenger is a liquid, it is believed that it works by two primary mechanisms to reduce the vapor phase H_2S levels. As the H_2S it generated by degradation of sulfur species (Figure 2) it is reacts with the H_2S scavenger and thus, no additional H_2S is evolved to the vapor phase. Assuming the asphalt is in a closed system; as the H_2S is consumed by the H_2S scavenger, the amount in the liquid phase decreases, as this happens, per the partitioning of the H_2S between the vapor and liquid, vapor phase H_2S will migrate back into the liquid phase were it can be consumed by the H_2S scavenger.

The H_2S levels of an asphalt sample from Location 1 with 5:1 vapor phase H_2S: H_2S scavenger dosage is shown in Figure 4. The plot in Figure 4 shows an average of three runs with the metal-based H_2S scavenger compared to a blank (no scavenger). The blank (no scavenger) data point used for this data set was not tested at 300° F to save on the limited amount of asphalt samples. The blank (no scavenger) was run first, with H_2S levels at 425 °F observed to be ~4000 ppm: thus an H_2S scavenger dosage of 800 ppm (5:1 vapor phase H_2S: H_2S scavenger) was used. To maintain consistency in the asphalt sample all the runs in the Figure 4 came from the same DOT-approved shipping can. In comparing the data at the 5:1 vapor phase H_2S: H_2S scavenger from Location 3 (Figure 3) and Location 1 (Figure 4), It can be seen that this dosage was nearly 100% effective for the sample in Location 3, yet only ~85% for the sample from Location 1 (see Figures 4 and 5). These two data sets reveal that in addition to having varying H_2S emissions profiles, H_2S scavengers perform differently in different asphalt samples. While the exact mechanisms for this varying performance would require a separate, independent study, it is believed that, as with H_2S emissions, varying sulfur species play a role, and as will be discussed, the viscosity of the sample.

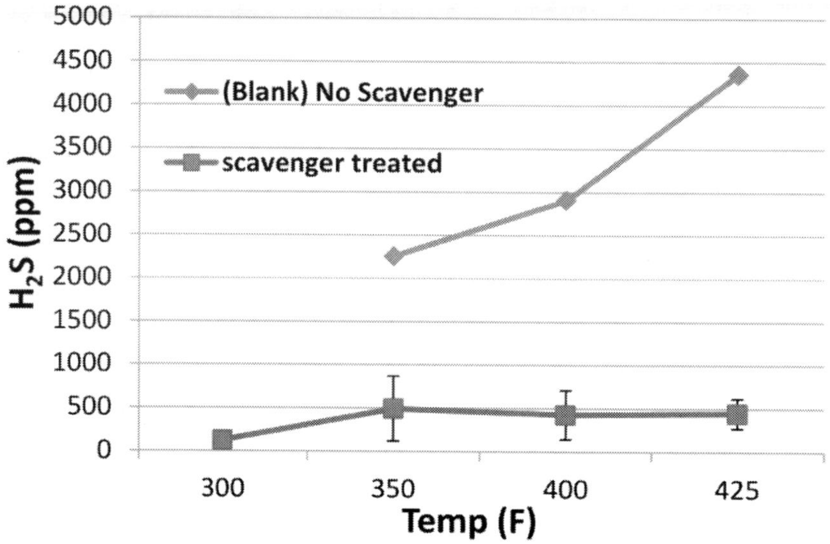

Figure 4: Headspace H_2S levels with 5:1 (vapor phase H_2S: H_2S scavenger) ratio

As can be seen in Figure 4, there is a flattening of the H_2S levels despite an increase in H_2S levels of the blank. As discussed above, It is believed that is due to scavenger molecules reacting with the H_2S as it is generated in the liquid phase. Figure 5 shows a plot of the data in Figure 4 in terms of percent effectiveness. For comparison, a plot with data from the Location 2 site was included; with a 5:1 vapor phase H_2S: metal-based H_2S scavenger.

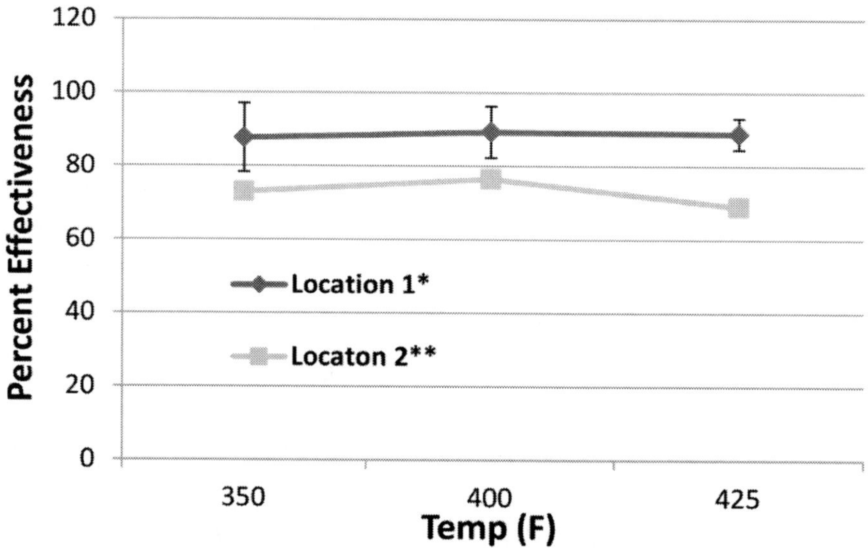

Figure 5: percent effectiveness of metal-based H_2S scavenger with asphalt samples from Location 1 and Location 2 *5:1 Vapor phase H_2S: Scavenger - average of three runs, ** 5:1 Vapor phase H_2S: Scavenger

In order to save on limited asphalt samples for the plot in Figure 5, the data point from Location 2 was run slightly differently. After being heated at 300° F an ASTM D5705 test was run to determine the H_2S levels and dosage rate as with the samples from Location 3 a 5:1 vapor phase H_2S: H_2S scavenger ratio was used. As Figures 4 and 5 demonstrate, the samples from Location 1 and sample from Location 2 follow a similar trend, with only a slight loss of percent effectiveness as the temperature increases, despite increased H_2S being evolved to the vapor phase due to cracking. The data presented in Figures 3-5 indicate that while solvent flashing may occur at temperatures as high as 425° F, it does not negatively impact the performance of the metal based H_2S scavenger.

Mixing/Driving Force

As discussed above, one of the objectives of this study was to determine the critical factors that determine H_2S scavenger performance. Figure 6 demonstrates the effects mixing can have on H_2S scavenger performance.

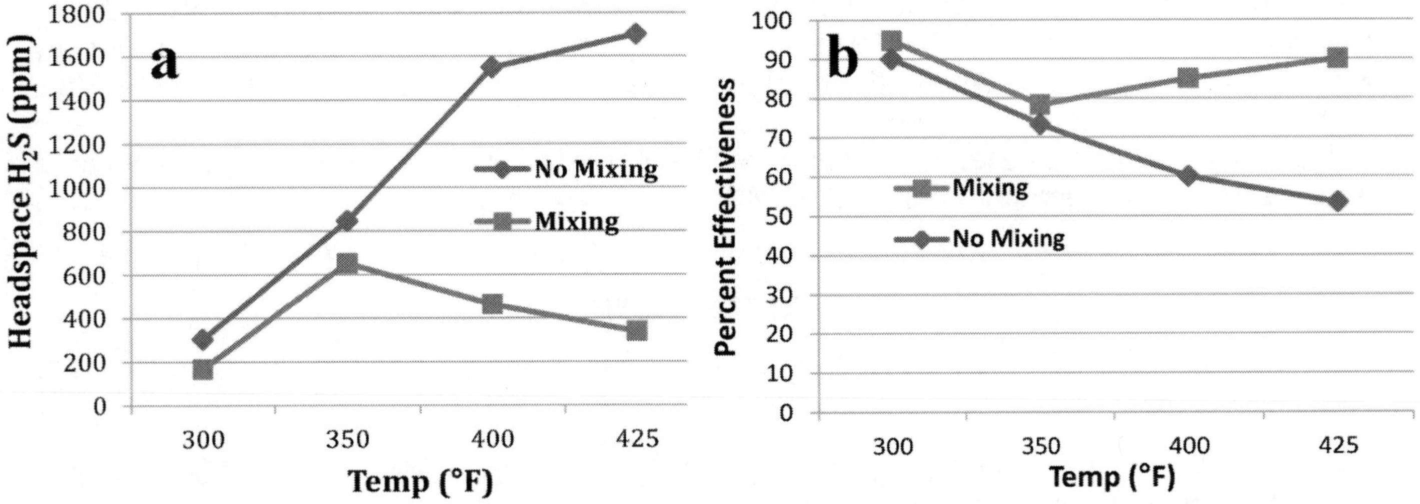

Figure 6: Mixing effects on H_2S scavenger performance in terms of a) H_2S levels and b) percent effectiveness

In Figure 6, a metal-based H_2S scavenger was used to treat an asphalt sample. For the no-mixing plot, the flask was first filled with 500 ml of asphalt and then the metal based H_2S scavenger was added without a stir bar used during the test. For the plot with mixing, the flask was pre-dosed with the metal based H_2S scavenger, followed by the addition of asphalt with a stir bar being used throughout the duration of the test. As seen in Figure 6, lack of mixing results in higher H_2S levels and renders the scavenger less effective, with vapor phase H_2S levels steadily increasing with temperature.

The data in Figure 6 demonstrate the critical importance of mixing and contact with H_2S scavenger performance. It is our hypothesis that as the additional H_2S is evolved into the vapor phase with increased temperature, the viscosity of the asphalt decreases and the mixing and contact efficiency of the H_2S scavenger improves resulting in better performance to account for the increased H_2S. For the H_2S scavenger to be effective, it must collide with the H_2S present in the sample. In the case of the higher temperatures, as the viscosity decreases, the H_2S and H_2S scavenger can move more readily through the fluid samples, resulting in more frequent collisions with H_2S and the improved efficacy of the H_2S scavenger.

CONCLUSIONS

The above data demonstrate that there are no adverse effects on the metal-based H_2S scavengers at temperatures up to 425° F. While it is likely that solvent flashing will occur at these temperatures, it does not adversely affect H_2S scavenger performance. The mixing of the H_2S scavenger in the asphalt improves due to the lower viscosity at higher temperatures, resulting in better performance that overcomes the additional H_2S generated due to degradation of sulfur species that form H_2S that migrates to the vapor phase. For an effective H_2S scavenger injection program, the temperature at which the H_2S scavenger is applied must be taken into consideration in order to have a better understanding of the amount of H_2S the will be in the vapor phase. In addition sufficient mixing and contact time are critical for effective H_2S scavenger utilization.

REFERENCES

1. J.Draper (2005), *Reactive Sulfur Mangement,* Reprinted From Hydrocarbon Engineering,
2. M.K. Amosa, I.A. Mohammand, S.A. Yaro (2010), *Sulphide Scavengers in Oil and Gas Industry – A Review*, NAFTA 61 (2) 85-92.
3. J.Draper, J. Stark (2012), *Abatement of Hydrogen Sulphide in Asphalt*, PTQ Q1.
4. D.C. Trumbore (Winter 1999), *Estimates of Air Emissions from Asphalt Storage Tanks and Truck Loading*, Environmental Progress, 18, 4: pp 250-259.
5. W.F. Faragher, J.C. Morrell, S. Comay (May 1928), *Thermal Decomposition of Organic Sulfur Compounds*, Industrial and Engineering Chemistry; pp 527- 532.
6. D.J. Slack, *Summary Report, Hydrogen Sulfide in Residual Fuel Oil and Storage Tank Vapor Space*, Texaco Refining and Marketing. Inc.

469626 Dynamic Simulation and Optimization for Automation of Distillation Columnshut-Down Operations

Abstract:

Dynamic Simulation and Optimization for Automation of Distillation Column

Shut-down Operations

Song Wang, Ziyuan Wang, Yiling Xu, and Qiang Xu

Dan F. Smith Department of Chemical Engineering

Lamar University, Beaumont, TX 77710, USA

Abstract

The shut-down operation of a distillation column is one of critical abnormal operations to a chemical/petrochemical plant. Normally, a planned distillation column shutdown operation consists of three operating procedures: (1) feed and liquid inventory reduction; (2) facility shut-down; (3) decommissioning including liquid discharge, vapor discharge, and nitrogen (N_2) purge. The shut-down operation involves concerns of operating time duration, on-spec product recovery, energy consumption, and emission generations, which needs sufficient cares. Because of the complexity in heat and mass transfer in different transient operating procedures, shut-down operations of distillation columns are usually conducted by manual and highly depending on industrial expertise in the reality. Considerably, there are many potential opportunities in the optimal operation of such shut-down operations.

In this paper, a methodology on dynamic simulation and optimization for automation of a distillation column shutdown has been developed. The shutdown operation of C_2splitter in an ethylene plant is employed in the base case study. The main objective is to accomplish a quick and automated shut-down operation on the basis of safety and in the meantime to save energy and materials. An Automated Selective Control (ASC) scheme is developed for improving the shutdown operation. The new PID controllers and different operational procedures are also investigated for obtaining a successful shutdown. In this work, associated with various shut-down control strategies, all their operating procedures during the column shutdown are programmed and automated in their rigorous simulation models, respectively. Through iterative dynamic simulations and fine tuning, the identified optimal operating case has significantly improved the column shutdown performance compared with the base case in terms of reducing the total shut-down time and recovering more on-spec products. Meanwhile, this study has also contributions on shut-down operating automation and smooth and safe transitional operations. It has laid out a solid foundation for distillation column shut-down automation and control improvement.

Keywords: Dynamic simulation, Process control, Shutdown, Distillation column, Ethylene plant, Automation.

463790 The Effect of Stirring Rates on Natural Gas Hydrates (NGH) Formation in Flow Reactor System: Experimental and Practical Approach

Abstract:

Abstract:

The liquefaction of natural gas (LNG) is one of the major modes of natural gas transportation between different regions. The liquefaction of natural gas requires cooling up to -161°C and the cost of liquefaction can range from 3 $ to 10 $ per MMBtu (millions of British thermal units). So unless there is a demand for higher LNG prices, the liquefaction cost could negatively affect LNG business and make liquefaction projects unprofitable.

In order to cut down liquefaction cost, the researchers are trying to study gas hydrates as a potential mode of natural gas transportation and storage in future. Gas hydrates are crystalline meta-stable compounds, made up of water and gas molecules. One volume of hydrate can store up to 180 volumes of natural gas at STP (standard temperature and pressure). Therefore, the development of hydrate production plants holds particular importance in future.

In this work, the formation rate of gas hydrates in a high pressure autoclave flow reactor system has been studied at different stirring rates using a magnetic stirrer. In the first phase, the system was first cooled from 20 ° C to 2 ° C at constant pressure for 18 hours. Then immediately in second phase, the system was left stable at 2 ° C for another 18 hours at constant stirring rate. The stirring rates were varied between 100-1400 Rpm (Rotations per minute) for each experiment and the gas hydrate crystal formation was observed by measuring pressure drop across the system during the second phase. The experimental results show that for a chosen system, the stirring rate does have an effect on the rate of gas hydrate crystal formation. The maximum gas hydrate crystal formation was observed within the stirring range of 550-750 RPM. It was found that for a chosen system there exists a threshold stirring limit, above or below which little or no hydrate crystal formation takes place. This fundamental study can be useful for modeling purposes and designing of a pilot scale hydrate production plant to study the economical and safety factor associated with natural gas hydrate storage and transportation.

Acknowledgement

This work was made possible by NPRP grant # 6-330-2-140 and GSRA # 2-1-0603-14012 from the Qatar National Research Fund (a member of Qatar Foundation). The statements made herein are solely the responsibility of the authors.

466172 Shalexenvironment: A Multi-Disciplinary Effort to Assess the Environmental Implications for Shale Gas Exploration Ad Production in Europe

Abstract:

In this presentation we will discuss the multi-disciplinary efforts implemented within the ShaleXenvironmenT consortium (http://shalexenvironment.org) to assess the potential impacts of shale gas exploration and production in Europe. The scientific activities implemented range from characterisation of shale core samples, quantification of the fluids contained in them, study of the fluid behaviour in nano- and meso-pores, design of fracturing fluids, optimisation of the water (both flowback and production) reclamation strategies, life-cycle assessment and comparison to other energy sources, as well as an analysis of socio-economical aspects related to the development of a shale gas industry in Europe.

463849 Study on Cold Energy Utilization System at Lcng Fueling Stations

Abstract:

Today the world energy market is in great transition and the demand of clean energy such as LNG increases rapidly. According to market prediction, the world demand for natural gas will increase about 2.1% yearly to 2030 and a higher proportion for Asian countries. In China, with the need of the environment protection and support from the government, the amount of natural gas vehicles is rapidly increasing, correspondingly with fueling stations. And it is known that the complete gasification of LNG (Liquefied Natural Gas) converting to CNG (Compressed Natural Gas) can return about 880kJ/kg of cold energy. If the cold energy recycle system is designed for station construction, it will bring not only economic but also environmental benefits. Nevertheless, comparing to the previous utilization of cold energy, key restrictions of utilization for LCNG fueling stations embody on three aspects. Firstly, discontinuous flow makes it hard to reuse the cold energy. It is necessary to convert it to a sustainable form such as reserved by PCM (phase change material). Secondly, CNG needs to be pressured as high as 25MPa where most heat exchangers in the market cannot sustain such pressure. A specific type of heat exchanger with matching flow pipe and measuring apparatus needs to be designed. Thirdly, considering the investment and maintenance cost, cold energy utilization system should be simple and compact. Besides, some other details are required to give sufficient consideration to make the whole system more economical and practical. Rare literature or patent relates to cold energy utilization in LCNG fueling stations. This paper will present an idea to make use of the cold energy and a set of apparatus to verify its feasibility.

Here are brief introduction of solutions to the cold energy utilization presented in this paper. Firstly, a suitable PCM to store the discontinuous cold energy of LCNG should guarantee safety, low toxicity and low flammability. At the same time, itâ€™s better to choose a common PCM to match downstream using cold energy apparatus. Thus, ice storage is a good choice satisfying the entire request above. On one hand, ice storage facilities are common in the market which can be bought and applied easily. On the other hand, ice/cold water can be used for air-conditioning in the LCNG stations. Secondly, traditional CNG fueling stations choose the vaporizer which exhausts cold energy directly to air. So a heat exchanger that could withhold high pressure up to 25MPa is quite important for the system. There are many types of heat exchanger such as plate, plate-fin, shell-and tube and so on. At present, according to the literature, only U-pipe and spiral wounded heat exchange can bear ultrahigh pressure. U-pipe ones are generally used in large scale system which needs more space. So spiral wounded heat exchanger with better exchange performance and less scale may be the suitable choice. Thirdly, besides the high pressure flow line and ice storage tank, a single-stage circulation with suitable intermediate refrigerating fluid would achieve the desired effect, which only involves one pump, one tank and pipe lines. Intermediate refrigerating fluid is divided into phase change material and non-phase change material and different material has different properties to affect the design of the size of heat exchanger and pipe diameter. In paper, investment cost, maintenance charge, and the cost payback period are discussed to make the whole system more attractive and practical.

Simulation cannot be avoided to make sure the feasibility of the system. HYSYS is a chemical engineering process dynamic simulation software, which allows designers to use the conceptual design and simplify the production process to complete the project work. Before the simulation, some fundamental parameters should be given combining references with experience. In consideration of safety in the lab, LNG is substituted by LN_2 (liquid nitrogen). EGS (ethylene glycol solution) which is chosen as the intermediate fluid should guarantee its fluidity when it is working at low temperature. The condensation point decreases while the mass concentration rises. When mass concentration reaches 70%, the condensation point could be the lowest point at -65â„ƒ. Considering the viscosity, 50% mass concentration of EGS is suitable. In addition, the values for the flow rate of LN_2, the efficiency of the pump, temperature, pressure and other parameters are explained in paper. By means of optimization by HYSYS, a series of process parameters of cold energy utilization system is given in details.

According to the design of cold energy utilization, a set of experimental apparatus is built to analyze the efficiency of the system. The whole equipment should approach to the real condition as much as possible which means the high pressure and discontinuous cold resource and limited working area. In the high pressure flow line, a made-to-order heat exchanger and pump make LN_2(liquid nitrogen) high up to 25MPa. And special venting pipe is designed to give off the N_2reducing the pressure and velocity without noise. Measuring transmitters connected to the data collecting computer are arranged in every inlet and outlet of the heat exchange device so that the amount of cold energy could be recorded. The equipment would achieve three different working conditions: ice-filling, ice-melting and ice-filling and melting at the same time. These conditions are based on the real demand in the air conditioning system in LCNG working space, which shows a good connection between cold energy recycle and utilization.

In the end, a discussion of the efficiency comparison between theory simulation and practical application demonstrates the feasibility of the cold energy utilization system and suggestions are made for further optimization and research.

key words: LCNG fueling stations, cold energy, ice storage, phase change material, experimental apparatus efficiency

463150 Techno-Economic Analysis of Shale Gas-to-Dimethyl Ether (DME) Process Via Direct Synthesis

Abstract:

Due to the volatile oil market and uncertainty in the crude oil supply in the future, it is necessary to develop cleaner and more carbon efficient non-petroleum based transportation fuels.[1] In the recent years, the use of dimethyl ether (DME) as an alternative transportation fuel is being strongly investigated. DME is used as a high quality fuel and has the potential to replace diesel due to its high cetane number (55-60).[2] DME can be produced by indirect and direct synthesis of syngas. The direct synthesis route has higher productivity compared to the indirect route but it has not been commercialized yet. Countries in the Asia-Pacific regions use coal to produce DME and are currently the largest consumers of DME. With low cost shale gas being abundantly available in the United States, DME can be produced from shale gas with less CO_2 emission and capital investment. Even though several studies have been conducted to developing catalysts and reactors for direct syngas-to-DME processes and plant-wide analysis of coal-to-DME processes, there is hardly any study in the open literature on a shale gas-to-DME process. With this motivation, we will present a techno-economic study of a shale gas-to-DME plant via direct synthesis using the process model developed in Aspen Plus® and economic model developed in Aspen Plus Economic Analyzer (APEA®).

In this work, a technically-feasible shale gas-to-DME process is synthesized first. Shale gas is first converted to syngas via pre-reformer and autothermal reformer and then converted to DME via direct synthesis. In the syngas production unit, a pre-reformer is required to convert the heavy hydrocarbons and prevent coking in the autothermal reformer. The raw syngas is sent to an acid gas removal unit to separate CO_2. The clean syngas is then sent to the DME production unit. The effluent of the DME reactor is then fed to a separation section where the unconverted syngas, DME, CO_2, water, and methanol are separated. A portion of the CO_2 recovered in the separation section is

utilized by recycling to the syngas production unit.

The presentation will focus on the following aspects: (1) development of kinetic models for pre-reformer, autothermal reformer, and DME synthesis reactor using the rate parameters regressed from experimental data, (2) development of a low cost novel separation system to separate unconverted syngas, DME, CO_2, water and methanol using DME as solvent, (3) development of the plant-wide model of a shale gas-to-DME process in Aspen Plus® and validation of plants section for which experimental data exist, (4) techno-economic analysis in APEA using the model developed in Aspen Plus®, (5) sensitivity study to analyze the effect of key design parameters on the main economic measures (i.e. net present value, internal rate of return, payback period), including design parameters such as steam to carbon ratio and oxygen to carbon ratio in the autothermal reformer, H_2/CO ratio in the DME synthesis reactor, technologies for acid gas removal (chemical or physical absorption), extent of CO_2 recycle, and utilization.

Reference

1. Azizi, Z.; Rezaeimanesh, M.; Tohidian, T.; Rahimpour, M. R. Dimethyl Ether: A Review of Technologies and Production Challenges. Chemical Engineering and Processing: Process Intensification. 2014, 82, 150â€“172.
2. Ogawa, T.; Inoue, N.; Shikada, T.; Inokoshi, O.; Ohno, Y. Direct Dimethyl Ether (DME) Synthesis from Natural Gas. Studies in Surface Science and Catalysis Natural Gas Conversion VII, Proceedings of the 7th Natural Gas Conversion Symposium. 2004, 379â€“384.

468659 Thermodynamic Modeling of Produced Water with Electrolyte NRTL Model: Aqueous Sr^{2+}- Na^{+} - so_4 $^{2-}$- Cl^{-} Quaternary System

Abstract:

Oil and gas activities especially hydraulic fracturing lead to production of significant amount of highly saline flowback produced water. The environmental concerns related to disposing the produced water as well as fresh water conservation strategy, encourage the reuse of the flowback produced water especially the direct reuse for hydraulic fracturing. However, scaling of some minerals at fracturing temperature and pressure is a serious concern for using saline water for fracturing. One of the leading cations in sulfate type scaling is strontium (Sr^{2+}) which indicates the importance of comprehensive thermodynamic modeling of aqueous system including Na^{+}-Cl^{-} ions which cause more than 90% of salinity of brine solution, and Sr^{2+}- SO_4^{2-}ions.

In prior studies, Pitzer model has been mostly used for saline water modeling. However, the model results have been shown to be accurate only for dilute solutions with ionic strength up to 6 molal. Further, the need for ternary interaction parameters in addition to binary interaction parameters makes the application of this model for multi-component produced water cumbersome. In most cases, there are not enough available experimental data to correlate numerous Pitzer model parameters.

Therefore in this study symmetric eNRTL activity coefficient model is used to model the aqueous Sr^{2+}-Na^{+}- SO_4^{2-}-Cl^{-} quaternary system. The model requires only molecule-electrolyte and electrolyte-electrolyte binary interaction parameters. The temperature dependency of these interaction parameters are described via Gibbs-Helmholtz equation including up to three temperature coefficients. Experimental data such as osmotic coefficient, mean ionic activity coefficient, vapor pressure, heat capacity, excess enthalpy, and salt solubility at different temperature up to 473.15 K and also at different electrolyte concentration up to saturation are used to identify the model parameters. The model provides accurate representation for all thermodynamic properties of the quaternary system.

ID# 469024

Ethane/Ethylene Liquefaction

Cutting Edge Processes to take the Fuel to the World

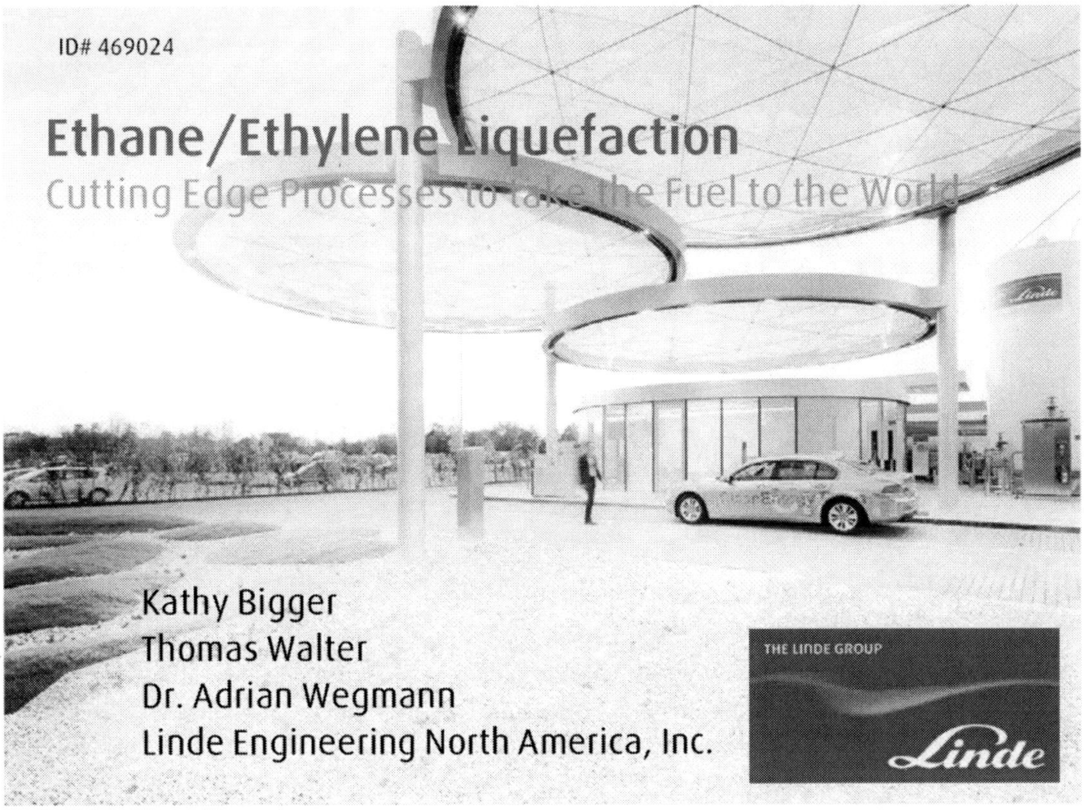

Kathy Bigger
Thomas Walter
Dr. Adrian Wegmann
Linde Engineering North America, Inc.

THE LINDE GROUP
Linde

Energy in the world

THE LINDE GROUP
Linde

- Total World Energy Consumption expected to grow 48% between 2012 and 2040
- Non OECD energy consumption to rise 71%, OECD energy consumption to rise 18%
- Economic growth (GDP) is key determinant for energy demand

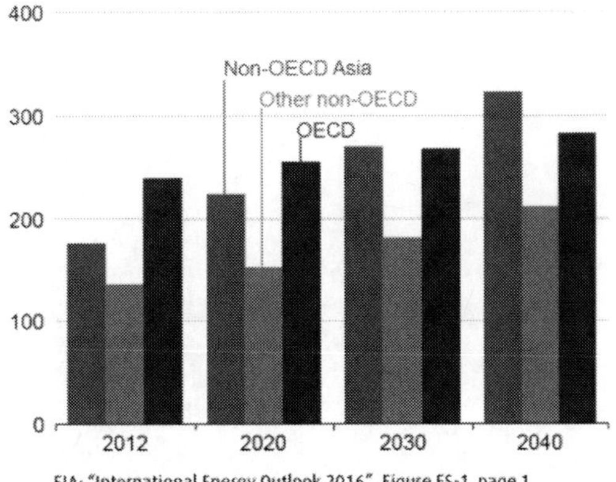

EIA: "International Energy Outlook 2016", Figure ES-1, page 1
World energy consumption by country grouping, 2012-40 (quadrillion BTU)

Energy in the world

- Growth per category:
  - Renewables
  - Nuclear
  - Natural gas
  - Liquid fuels
  - Coal

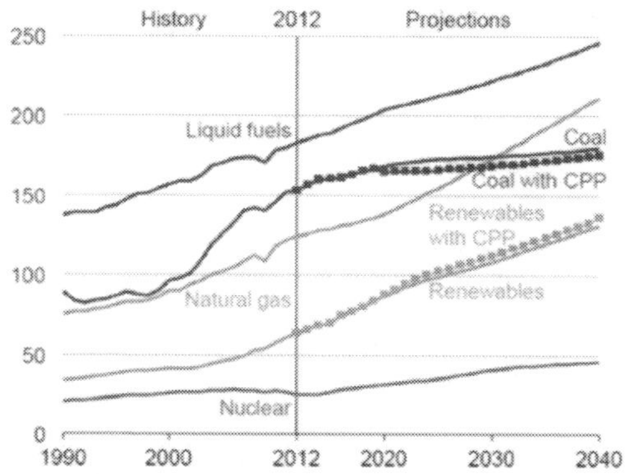

Note: Dotted lines for coal and renewables show projected effects of the U.S. Clean Power Plan.

EIA: "International Energy Outlook 2016", Figure ES-2, page 1
Total world energy consumption by energy source, 1990-2040 (quadrillion BTU)

Natural gas is projected to be the fastest-growing fossil fuel!

Natural gas – sources

- Ethane is a byproduct of the natural gas production
  - Shale gas
  - Tight gas
  - Associated gas
  - Coalbed methane
  - Other natural gas

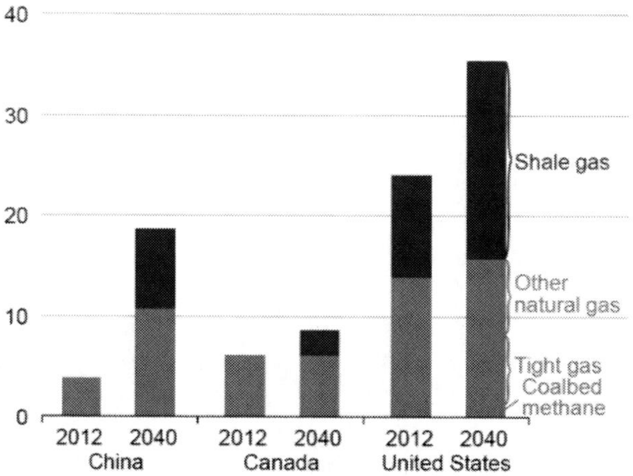

EIA: "International Energy Outlook 2016", Figure 3-3, page 38
Natural gas production by type, 2012 and 2040 (trillion cubic feet)

Investment in ethane/ethylene infrastructure

THE LINDE GROUP Linde

- Driven by abundant availability
  - Current surplus in the United States: ~500'000 bbl/day
  - Expected surplus in the United States in 2020: ~150'000 bbl/day
- Required by geographical mismatch between source and consumer
- Ethane Cracker capacity is being expanded due to cost advantage over naphtha crackers
  - Installed ethylene capacity in the USA in 2012: ~24 million tons per year
  - Anticipated capacity in 2020: ~35 million tons per year
- Challenges:
  - Lack of dedicated pipeline system
  - Physical properties require highly specialized tank equipment similar to LNG
- Solutions:
  - Liquefaction
  - Processing capabilities in unusual locations (Badlands NGL Polyethylene plant in North Dakota)

Ethane export

THE LINDE GROUP Linde

- Export
  - Pipeline to Canada since late 2013
  - Marcus Hook, Pennsylvania: 35'000 bbl/d capacity since March 2016
  - Morgan's Point, Texas: 200'000 bbl/d capacity since September 2016

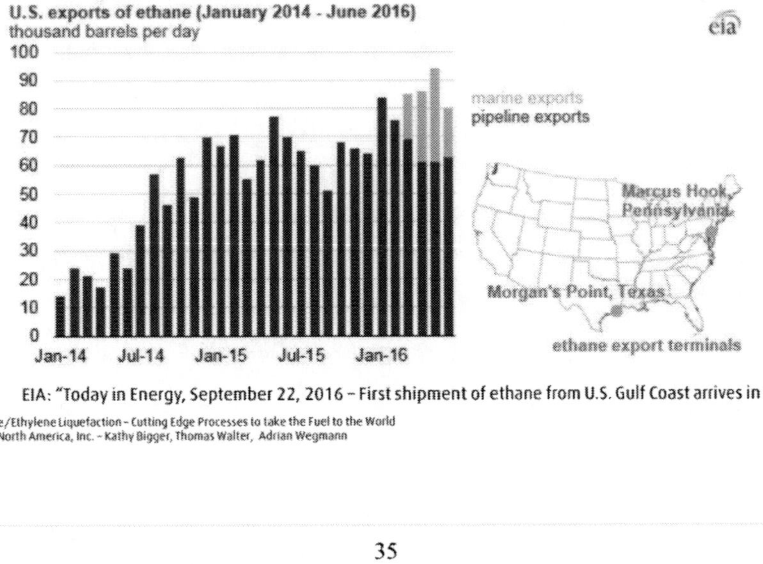

EIA: "Today in Energy, September 22, 2016 – First shipment of ethane from U.S. Gulf Coast arrives in Europe"

Drivers for finding the optimized solution

THE LINDE GROUP
Linde

- Customer desires
  - OPEX – Compressor power, cost of refrigerant
  - CAPEX – Equipment count & size
  - Compliance implications, legal requirements
- Limiting factors
  - Physical size of plate fin heat exchangers (PFHX)
  - Refrigerant compressor power
- General dependencies
- Range of feasible solutions

Parameter study

THE LINDE GROUP
Linde

- Heat exchanger size – number of cores
- Piping arrangement
- Coldbox size
- Process conditions (pressure determines vaporization temperature)

– 30'000 variations of process setup

Pareto Frontier

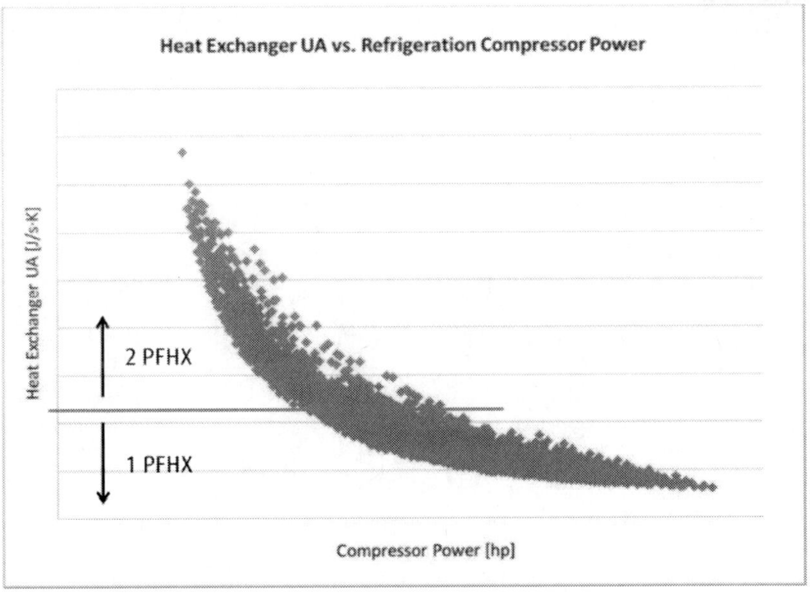

Conventional liquefaction process
Open Loop

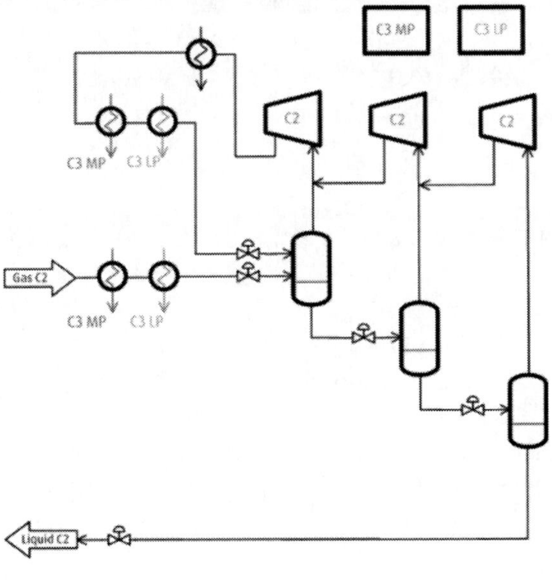

Open loop Process

- Conventional process utilizing STHE and block in shell to liquefy the C2
- Typical inlet pressure from pipeline >60 bar (900psi)
- MP and LP propane or propylene refrigeration needed to precool the C2
- Liquefaction by multi-stage flash and recompression of the flash gas
- Conventional process can also be designed as a closed loop resulting in increased power consumption

Mixed refrigerant liquefaction process
LIMUM®1

THE LINDE GROUP
Linde

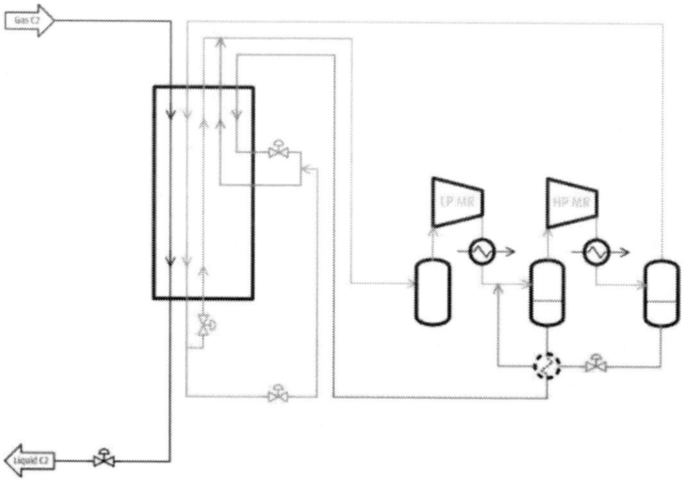

Single Cycle Process

- Typical for peak shaving or small scale LNG plants; process adapted to liquefaction of C2
- Simple process, low equipment count
- One refrigerant cycle providing refrigeration duty
- Pre-cooling, liquefaction and sub-cooling cycle realized with a plate-fin heat exchangers
- Lower specific energy consumption compared to an expander process

StarLNG™
Liquefaction of CH_4 vs. C_2H_6 / C_2H_4

THE LINDE GROUP
Linde

@1.013 bar/ 14.7 psi	CH_4	C_2H_6	C_2H_4
Liquefaction temperature	-161.6°C (-258.9°F)	-88.7°C (-127.7°F)	-104.1°C (-155.4°F)
Heat of vaporization	512.2 kJ/kg (220.2 Btu/lb)	487.5 kJ/kg (205.7 Btu/lb)	476.2 kJ/kg (204.7 Btu/lb)

@40 bar/ 580 psi	CH_4	C_2H_6	C_2H_4
Liquefaction temperature	-87.3°C (-125.2°F)	22.6°C (72.6°F)	-1.1°C (29.9°F)
Heat of vaporization	160.7 kJ/kg (69.1 Btu/lb)	171.6 kJ/kg (73.8 Btu/lb)	179.2 kJ/kg (77.0 Btu/lb)

The heat of vaporization is similar, but the liquefaction temperature is significantly different.
Rule of thumb: The specific energy consumption compared to LNG is approx. 5.6 times smaller for Ethane and 4.1 times smaller for Ethylene!

Comparison energy consumption and equipment count
Mixed refrigerant vs. conventional

	Mixed refrigerant LIMUM®1	Conventional Open loop
Relative power consumption	**1.0** 1), 2)	**0.97** 1), 2)
Heat exchanger	2x Air Cooler/Shell&Tube 1x Plate-fin 1x Shell&Tube (optional)	3x Air Cooler/Shell&Tube 3x Shell&Tube 1x Block in Shell
Compressor	2x Mixed Refrigerant stages	3x C2 Compressor stages 2x C3 Compressor stages
Separators	3x Suction/Discharge Vessel	5x Suction/Discharge Vessel
Make-up	2-3x Storage CH_4, C3(opt.) and C4	1x Storage C3
Total 3)	**10-12 Equipment**	**18 Equipment**

1) The power consumption for an Ethylene liquefaction unit was 94.4 vs. 91.3 kWh/t_{C2}
2) Power consumption strongly dependent on process conditions, cooling water vs. air cooler, etc.
3) Liquefaction unit only. Equipment for Dehydration, BOG and Storage is identical

ID# 469024: Ethane/Ethylene Liquefaction - Cutting Edge Processes to take the Fuel to the World
Linde Engineering North America, Inc. - Kathy Bigger, Thomas Walter, Adrian Wegmann

13

Summary

- Ethane is abundantly available in the US for the foreseeable future

- Ethylene production is forecast to substantially increase in the near term

- Both ethane and ethylene are considerably easier to liquefy than methane

- There is a large potential for the US to become a supplier of both ethane and ethylene to nations without any such resources readily available

- Modern, optimized technical concepts can provide sustainable solutions for the ever more challenging energy landscape on our planet!

ID# 469024: Ethane/Ethylene Liquefaction - Cutting Edge Processes to take the Fuel to the World
Linde Engineering North America, Inc. - Kathy Bigger, Thomas Walter, Adrian Wegmann

14

THE LINDE GROUP

Linde

Thanks for your attention.

463765 Absorption of Mixed Refrigerant in Lubricant Oil

Abstract:

Absorption of mixed refrigerant in lubricant oil

Yang,Song Wensheng,Lin

Institute of Refrigeration and Cryogenics, Shanghai Jiao Tong University,P.R.China

Abstract

Of all processes producing liquefied natural gas (LNG), mixed refrigerant cycle is considerably appealing for its high efficiency. In a small scale mixed refrigerant cycle, utilizing injected screw compressors makes the whole compression unit more compact. However mixed refrigerant may dissolve in lubricant oil existing in injected screw compressors. It is pretty essential to understand the detailed knowledge of mixed refrigerant absorption in refrigerant-lubricant oil mixtures, as the change in compositions of mixed refrigerant can directly influence some important operation parameters such as process energy consumption, heat exchangers performance, etc.

To determine the optimized components of mixed refrigerant in single mixed refrigerant (SMR) liquefaction process, without considering the absorption of refrigerant in oil, the Knowledge Based Optimization (KBO) method was adopted. During this optimization process, there were two limiting conditions for operation: the minimum heat transfer temperature difference in the LNG heat exchanger was set 3K and liquid entering compressors was not allowed. The P-R equation of state, which is suitable to calculate the hydrocarbon mixtures in particular, was adopted in the simulation. Typically, energy consumption of unite LNG product directly affects the running cost of the liquefaction system, so the minimum specific power consumption was set as objective function. Then refrigerant components in SMR LNG process were optimized for both four and five compositions by adjusting sub-flow of the components from low to high boiling point according to the KBO method. At the end of the optimization process, the optimal constitution was gained with methane 21%, ethylene 14%, propane 58%, nitrogen 7% in the four-composition mixed refrigerant and methane 19%, ethylene 30%, propane 4%, i-butane 39% and nitrogen 8% respectively. These optimal results were then used as basis of follow-up studies on the effect of single refrigerant component change on operation parameters as well as experimental investigation on refrigerant absorption in lubricant oil.

After the optimal constitution of mixed refrigerant was gained, composition change due to dissolving in lubricant oil was considered. In order to investigate the impact of the change of each component on the operation parameters referred to above, a variable-controlling approach based on HYSYS was adopted. More precisely, the influence of methane, ethylene, propane and isobutene was investigated respectively. Firstly, we investigated the effect of each component change on the specific power consumption. Different component reducing affect specific power consumption differently due to their thermophysical property difference. To study this effect more specifically, the content of each component was reduced gradually, each time 0.01kmol/h, while other components remained the same and the specific power consumption was recorded at the same time. Methane, ethylene, propane, and i-butane was individually operated in the same way until the minimum heat transfer temperature difference in the LNG heat exchanger was less than 3K. During this process, we can obtain that with the decrease of each kind of components, the specific power consumption declined in a linear trend. However, reducing the same amount of each component, the specific power consumption reduced in a different degreeâ€"reducing methane had the greatest impact, followed by ethylene, propane in the four-composition refrigerant process and the five-composition refrigerant shared the same trend with i-butane had the smallest impact. It is noteworthy that, nitrogen is hardly soluble in lubricant oil, so we did not consider its reduction. Secondly, the effect of each component change on the minimum temperature difference in LNG heat exchanger was investigated. Each component was reduce in the same way referred to above, during this process the minimum temperature difference was recorded. It can be obtained that in the four-composition refrigerant process, reducing propane impacted to the greatest extent, followed by methane and ethylene. However, the five-composition refrigerant process saw a different pattern with i-butane impacted most, followed by propane, ethylene and methane. This trend can be explained by temperature sensitive range of each component.

As well as single component analysis based on HYSYS, experiments focusing on overall absorption of mixed refrigerant in lubricant oil were also conducted.During this experiment process, we studied four-component gas, which was constituted by 21% methane, 14% ethylene, 58%propane, 7%nitrogen, entering lubricant oil. The model number and brand of lubricant oil were determined by actual LNG process. In this experiment, we tested three different kinds of lubricant oil (CP-1010-100, CP-1005-100, CP-1516-100) which were widely used in the SMR process compressors. In order to ensure the lubricant was sufficient, the tank was designed to be 10 liters.

Initially, the tank was evacuated, after which lubricant oil was added to the desired weight (10kg). Then turn on the heater until the oil reached 303K by regulating the regulator. After that, open the valve of the high pressure gas bottle, then the mixed gas was introduced into the lubricant oil. Regulating the valve to make sure the pressure in the tank was 0.5MPa initially. Finally, the component of the exhaust mixed gas was measured by gas chromatograph. When the component measured kept unchanged for two times, wrote down the specific percentage of each component and mixed gas mass flow of intake and exhaust. Apart from that, the instantaneous system pressure and temperatures in vapor and liquid regions were both recorded at temperatures of 303K, 313K, 323K, 333K, 343K and 353K. When the mixed gas in the bottle was used up, the lubricant oil should be discharged and the tank was cleaned for the next kind of lubricant oil. Because nitrogen is hardly soluble in lubricant oil, its mass flow kept unchanged, which could be used as a leak detection during the experiment process.

The mass of each component of mixed refrigerant absorbed in lubricant oil was calculated by the experiment data. When the calculation was completed, the results were then used to investigate the impact of mixed refrigerant absorbed in lubricant oil on operation parameters change in the LNG liquefied process through HYSYS.

466240 The Steam Iron Process for Pressurized Hydrogen Production

Abstract:

The commitment to fuel cell powered electric vehicles offer a carbon emission free mobility in the near future. A key requirement to achieve this goal is the supply of low cost hydrogen produced out of renewable resources. Currently more than 90% of the hydrogen production involves the utilization of fossil fuels in large scale centralized plants. The supply of cost-competitive hydrogen by renewable resources requires a different production infrastructure. Transportation distances of resources and of hydrogen itself have to be minimized and the production has to be realized in decentralized small scale production facilities. At Graz University of Technology the reformer steam iron process is developed as a scalable, on-site process, to emphasize on the requirements of natural resources. The process combines the steam reforming with the steam iron process (a chemical looping water splitting process) in one single unit.

In the first step of the process a syngas is generated by catalytic steam reforming of hydrocarbons according to eq.1. $C_xH_y + xH_2O$ -> $(x+y/2)H_2 + xCO$ (1) This syngas is directly used for the reduction of an oxygen carrier (eq. 2) which is a mixture of hematite (Fe_2O_3) and supporting metal oxides, prepared by mechanical mixing. $Fe_2O_3 + 3H_2/CO$ -> $2Fe + 3H_2O/CO_2$ (2) The consecutive oxidation of iron with steam leads to the formation of magnetite (Fe_3O_4) and pure hydrogen (eq. 3), an additional oxidation step with air regenerates the magnetite back to its initial hematite state (eq. 4). $3Fe + 4H_2O$ -> $Fe_3O_4 + 4H_2$ (3)

$2Fe_3O_4 + 1/2O_2$ -> $3Fe_2O_3$ (4) In the reformer steam iron process the reduction is performed at ambient pressure to utilize a high fuel gas conversion efficiency due to the favorable reaction equilibrium. The oxidation on the other hand can be executed at an increased pressure to yield compressed pure hydrogen without additional gas compressors, which improves the efficiency of the system.

Isothermal cyclic pressure experiments at 750 °C were performed in a lab scale test rig as a proof of concept for pure pressurized hydrogen production. The goal of the experiments was to investigate (i) possible draw-backs of the production of compressed hydrogen by stability constrains of the oxygen carrier or operational limitations, (ii) influences of an increased system pressure on the hydrogen purity regarding carbon contaminations, (iii) the characterization and quantification of the contaminations and (iv) the analysis of the oxygen carrier conversion regarding efficiency and stability.

The reduction reactions were performed at ambient pressure with hydrogen or a synthesis gas mixture of $H_2/CO/CO_2$. The re-oxidation of the oxygen carrier at an increased pressure was achieved by pumping liquid water into the system which is evaporated in the heated inlet. The steam oxidizes the oxygen carrier by releasing hydrogen. The pressure build-up is realized by closing the out-going valve after the cooling system. In this case the unreacted steam is condensed and the pressure increase can be directly related to the amount of produced hydrogen.

Â A successful hydrogen production up to 55 bar was achieved in the experiments with a hydrogen purity to fit the requirements of low temperature fuel cells. The oxygen carrier showed signs of structural degradation with no evident impact of the different pressure conditions in the prior oxidations. The linear pressure increase in the ï¬rst part of each oxidation maintained at a stable level throughout the entire investigation thus the oxidation rate was not reduced by the solid degradation. Sintering was identified as the source for the loss of cycle stability which is mainly inducted by the high reaction temperatures of 750 Â°C. The oxidations indicated no visible influence on the purity by the different system pressures. Thus the process is suitable for the direct production of pressurized hydrogen using hydrocarbons as feed.

Acknowledgements

This work is funded by the IEA-Research Cooperation on behalf of the Austrian Federal Ministry of Transport, Innovation and Technology. The authors gratefully acknowledge support from NAWI Graz.

473113 Integrated Thermochemical Water-Splitting Reactor Fuel Cell System for Electricity Generation

Abstract:

A continuous thermochemical water-splitting reactor for hydrogen generation has been developed and integrated with a fuel cell for electricity generation. To perform a water-splitting reaction, ferrite materials were synthesized inside a porous yttria-stabilized zirconia (YSZ) substrate using a sol-gel technique. These materials were analyzed by scanning electron microscopy, x-ray diffraction and BET surface area analyzer. In particular, Li-based ferrite materials were immobilized onto a porous YSZ substrate and loaded inside a continuous water-splitting reactor where both regeneration and water-splitting steps were performed at lower temperatures. Hydrogen and oxygen were continuously monitored and recorded using the online sensors interfaced with a computer. The outlet from a continuous thermochemical water-splitting reactor is integrated with a hydrogen fuel cell module. The fuel cell current was continuously recorded and analyzed at different load conditions. The power output and fuel cell or stack current were analyzed at different hydrogen production rates. The results obtained on the characterization of ferrite materials, hydrogen volume generation and the electricity generation will be presented.

466622 The Development of Microchannel Reactors for Hydrogen Production: Experimentation and CFD Modelling

Abstract:

A microchannel reactor for hydrogen production from ammonia is simulated with a three-dimensional (3D) CFD model using Comsol Multiphysics, in order to obtain a mathematical framework to understand the reaction-coupled transport phenomena associated with the microchannel reactor. The catalysts used consisted of commercial Ni-Pt/Al_2O_3 and Ru/Al_2O_3 catalysts in the form of a porous wash-coat deposited on the walls of the micro channels. The transport processes and reactor performance are elucidated in terms of velocity, temperature, and species concentration distributions, as well as local reaction rate and ammonia conversion profiles. The baseline case is first investigated to determine the behaviour of the microchannel reactor, then microstructural design and operating parameters are methodically altered around the baseline conditions to explore the optimum values. The simulation results show that an optimum ammonia space velocity (GHSV) yields 99.1% ammonia conversion and a power density of 32 $kW_e\ L^{-1}$ at the highest operating temperature of 973 K for the Ni-Pt based catalyst. It is shown that a 40-Î¼m-thick porous wash coat is most desirable at these optimum conditions. Also a small channel hydraulic diameter (225 Âµm) is observed to contribute to high ammonia conversion. It is reported that the Ru/Al_2O_3 gives a superior yield than the Ni-Pt/Al_2O_3 at optimal conditions and that experimental and model results agree very well. The results obtained confirm that a catalytic microchannel reactor can be most desirable for small scale hydrogen production from ammonia, required especially for the operation of fuel cells.

458104 Thermodynamic Modeling of the Hybrid Sulfur (HyS) Cycle for Hydrogen Production

Abstract:

Abstract

The hybrid sulfur (HyS) cycle has received much attention over the past few decades as a possible means of efficient large-scale production of clean hydrogen from high-temperature nuclear or solar heat sources [1]. A typical HyS cycle consists of a thermochemical step and an electrochemical step, the net result of which is the splitting of water into hydrogen and oxygen. In order to fully understand the limiting HyS heat and electric power requirements and to optimize the operating and design parameters, it is important to perform a thermodynamic analysis of the highly nonideal aqueous SO_2-H_2SO_4 system. Therefore, we develop a comprehensive thermodynamic model for the sulfuric acid-sulfur dioxide-water ternary system based on the symmetric electrolyte NRTL (eNRTL) [2] activity coefficient model framework in conjunction with the PC-SAFT [3] equation-of-state. This model is an extension of the previously published thermodynamic model of the sulfuric acid-water [4] binary system and additionally considers one step hydration of sulfur dioxide molecules. The eNRTL binary interaction parameters and the chemical equilibrium constants are regressed from the experimental SO_2solubility data in aqueous sulfuric acid solutions and the ternary model is validated over a wide temperature range from 283.15 to 383.15 K, and acid concentrations from pure water to pure acid. The model should be very useful in supporting process research, development, and design of the advanced direct solar water splitting process based on the HyS cycle.

Keywords:

Hybrid-Sulfur (HyS) cycle, Electrolyte NRTL, Hydrogen, SO_2 solubility, Sulfuric Acid

References:

[1] Gorensek MB, Summers WA, "Hybrid sulfur flowsheets using PEM electrolysis and a bayonet decomposition reactor," *Int. J. Hydrogen Energy*(2009) 34, 4097-4114.

[2] Song Y, Chen C-C, "Symmetric electrolyte nonrandom two-liquid activity coefficient model," *Industrial & Engineering Chemistry Research*(2009) 48, 7788–7797.

[3] Gross J, Sadowski, G, "Application of the Perturbed-Chain SAFT Equation of State to Associating Systems," *Ind. Eng. Chem. Res.* (2002) 41, 5510-5515.

[4] Que H, Song Y, Chen C-C, "Thermodynamic Modeling of the Sulfuric Acid-Water-Sulfur Trioxide System with the Symmetric Electrolyte NRTL Model," *J Chem Eng Data* (2011) 56, 963-977.

471842 Utility Perspectives on the Hydrogen Economy

Abstract:

Hydrogen is poised to become a preferred energy carrier in a deeply decarbonized energy future. While today's utilities make, move, or sell today's energy carriers of electricity and natural gas, future utilities may include hydrogen. This talk will focus on potential roles for today's utilities as they contemplate energy carriers of the future. It includes a survey of hydrogen production, not only from conventional steam methane reforming or coal gasification, but also technologies such as advanced electrolysis, methane thermolysis, water photolysis, and thermochemical hydrogen production cycles. It will discuss hydrogen delivery, a potential future of spatial and transient energy storage using hydrogen, the emergence of a hydrogen economy, and roles for hydrogen to provide clean energy in the near-term. Finally, the hydrogen economy will be contrasted against alternative future scenarios.

461264 Convective Heat Supply Reactor Concepts for the High Temperature Pyrolysis of Methane

Abstract:

Contrary to the popular belief, fossil energy consumption is not limited by its availability but is constrained by environmental considerations. Over the past decades, the world's natural gas proven reserves have been increasing steadily, from 2592 trillion cubic feet in 1980 to 6972 trillion cubic feet in 2014 [1], and if non-conventional sources like methane hydrates or shale gas are included, this increase is even steeper. Many scenarios for the future production of energy without CO_2-emissions have been investigated. The consensus seems to be directed towards renewable energy; however, a complete transition to renewable energy is not foreseeable in the near future [2]. It is thus necessary to search for bridging technologies. Among these, hydrogen presents itself as a very attractive alternative energy carrier due to the possibility of using existing natural gas infrastructure as well as for its facile integration in fuel cells and other clean energy technologies.

Hydrogen can be generated with low to zero CO_2-emissions by pyrolysing hydrocarbons at high temperatures. The pyrolysis of methane can be described by the reaction:

$$CH_4 \rightleftarrows C(s) + 2H_2$$

CH4⇄Cs+2H2

$$\Delta H_{R,1000°C} = 91.52\ kJmol^{-1}$$

DHR,1000°C=91.52 kJmol-1

DGR,1000°C=-49.28 kJmol-1

$$\Delta G_{R,1000°C} = -49.28\ kJmol^{-1}$$

This reaction is endothermic and becomes thermodynamically spontaneous above 820 K and kinetically feasible at temperatures around 1073 K. The reaction can be catalysed at lower temperatures, but the carbon deposition on the catalyst and the limited equilibrium conversions leading to a separation problem and recycle streams are still issues that have to be addressed. The high temperature pyrolysis remains an alternative, but is not without challenges: carbon deposition is still a major hurdle, which can lead to blockage of the reactor, and the endothermicity of the reaction necessitates the introduction of heat at higher temperatures.

Convective heat supply reactor concepts offer advantages that can help circumvent both problems. In this work, three concepts are presented:

The first two concepts are based upon the use of liquid molten media, such as molten metals. On the one hand, the liquid medium acts as a barrier preventing carbon deposition at the heat transfer surface, as well as a heat transfer material due to its excellent transport properties. The use of molten metal bubble columns was first proposed by Steinberg [3]. A modified version was explored by Serban et al. [4] exhibiting a relatively high conversion (57%) for low temperature operation (750°C). In this publication, carbon deposition occurred at both the reactor surface and in the sparger used to feed the methane. Currently a bubble column reactor with a fixed-bed is being investigated by the group of Prof. Abanades in the IASS in Postdam in cooperation with the Karlsruhe Institute of Technology amongst others [4]. Their results show a hydrogen yield of 30% at 1000°C with little carbon deposition on the reactor surface. However, the lack of control over the residence times in bubble columns inherently leads to a decrease in conversion. To overcome this difficulty, the use of a capillary reactor using molten metals in slug flow regime (Fig. 1) was studied at temperatures up to 1100°C [5], showing promising results, although only relatively low conversions (30% methane) have been achieved so far.

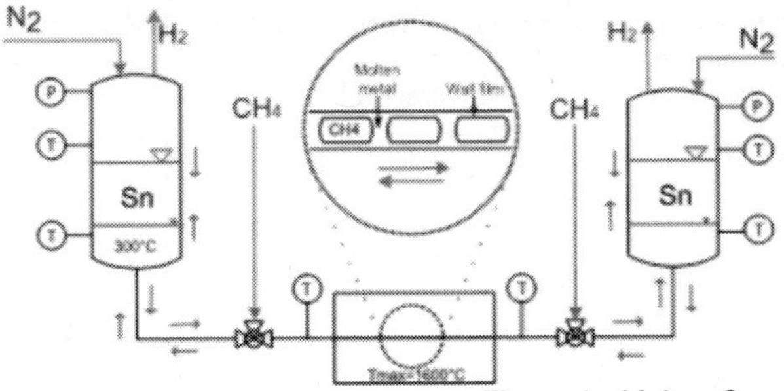

Figure 1. Molten Sn capillary reactor in slug flow regime

To increase conversions, working at still higher temperatures is the logical approach when the reactor length is limited. However, initial experimental results (Fig. 2) indicate increased carbon deposition, which can be explained by the increased rates of carbon formation, coupled with insufficient wetting of the heat transfer area by the molten metal as well as morphological differences in the carbon formed. To resolve this problem, two techniques can be used: the first one is to tweak the wetting properties of the molten metal, to avoid carbon deposition. A second approach would be to modify the reactor design appropriately to guarantee film formation while maintaining the advantages of slug flow operation.

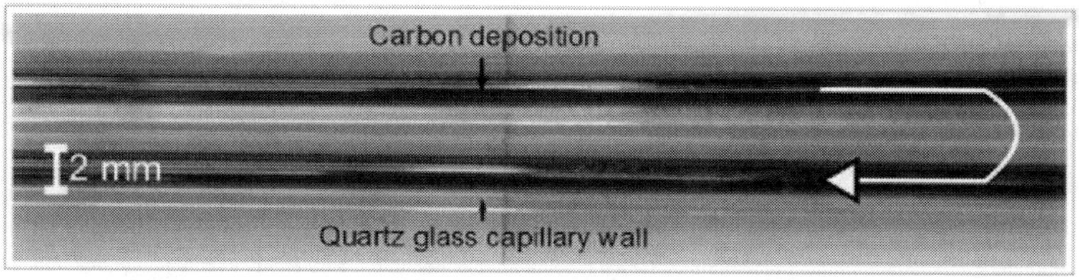

Figure 2. Carbon deposition in a capillary reactor at 1300°C

In this work, the experimental results for the first approach are presented, where the wetting properties of the melt were modified using a different alloy (Ga-In-Sn) rather than just tin. For the second approach, a theoretical study on a countercurrent falling-film reactor (Fig. 3) is presented. The falling-film reactor uses a similar concept to the capillary reactor, employing the molten metal as both a heat transfer medium and to protect the heat transfer surface, thus retaining the advantages of the slug flow regime. In addition, countercurrent operation inherently prevents the occurrence of the back reaction which could potentially take place in the slug flow regime. Furthermore, the residence time of the gas although not as well-defined as for the slug flow regime, can be still considered narrow enough to assure plug flow. The mathematical modelling and the simulation results are presented together with preliminary experimental results regarding film stability.

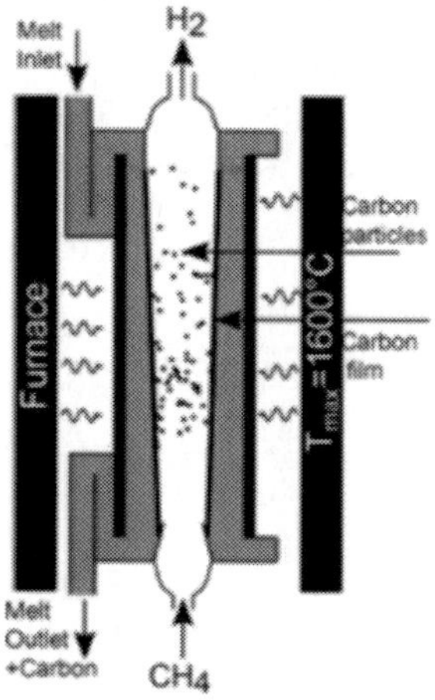

Figure 3. Falling molten Sn film reactor.

Finally, the third concept uses gas as a convective heat carrier. Previous work at our laboratory used a porous-wall reactor [6], where the carrier gas flowed through a porous wall into the reaction zone, avoiding the carbon deposition on the outer walls. However, methane diffusion through the membrane proved to be higher than anticipated, and as a result carbon deposits were still found in the outer region. To avoid this diffusion, a side-jet reactor (Fig. 4) concept was developed, where the carrier gas flows through small holes in the inner tube generating a greater gas velocity in comparison to the porous wall. This minimizes the rate of diffusion of methane into the outer zone and provides a better mixing with the methane and thus an improved heat transfer. In this work, the modelling of the side-jet reactor is presented as are the preliminary experimental results.

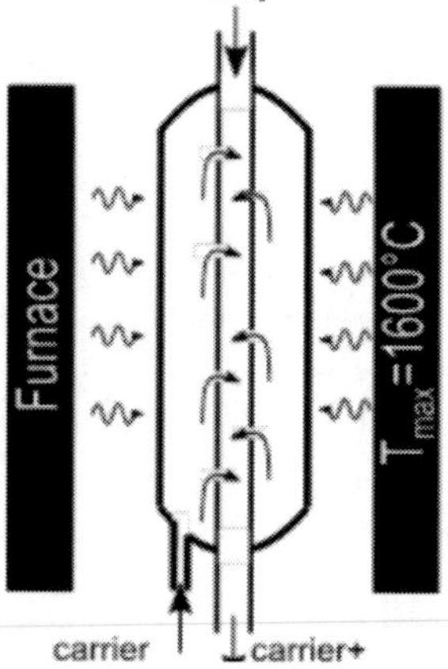

Figure 4. Side-jet reactor concept.

Acknowledgements

Financial support from the German Research Council DFG (Grant No. AG 26/15-1) is gratefully acknowledged.

References

[1] U.S. Energy Information Administration. International Energy Statistics. Retrieved from: http://www.eia.gov/cfapps /ipdbproject/iedindex3.cfm?tid=3&pid=3&aid=6&cid=regions&syid=1980&eyid=2015&unit=TCF on October 2015.

[2] G. Kreysa, DGMK Conference "The Future Role of Hydrogen in Petrochemistry and Energy Supply", Berlin, Germany, October, 2010.

[3] M. Steinberg, Int. J. Hydrogen Energy. 24(8) (1999) 771-777.

[4] M. Serban, M. A. Lewis, C. L. Marshall, R. D. Doctor, Energy Fuels. 17 (3) (2003) 705-713

[5] T. Geißler , M. Plevan , A. Abanades, A. Heinzel, K. Mehravaran, R.K. Rathnam, C. Rubbia, D. Salmieri, L. Stoppel, S. Stückrad, A. Weisenburger, H. Wenninger, T. Wetzel, Int. J. Hydrogen Energy. 40 (41) (2015) 14134-14146.

[6] I. Schulz, D.W. Agar, Int. J. Hydrogen Energy. 40(21) (2015) 11422-11427

[7] Fabian Goebel, Bachelor Thesis, Technische Universität Dortmund, 2015.

471254 Study of Bio-Inspired Nanostructured Materials for Solar Energy Conversion

Abstract:

Solar energy conversion has been drawn a great interest in our scientific community due to the demand for renewable energy. The necessary steps to improve solar energy conversion, including photovoltaics and water splitting, are solar energy harnessing, charge separations, and catalytic reactions. Nature always inspires us to mimic natural photosynthetic systems such as, Purple photosynthetic bacteria, revealing a new light harvesting protein with efficient channels for energy and electron transport.

This study focuses on developing artificial photosynthetic systems for understanding the fundamentals of energy transfer, and electron transport in both water splitting, and photovoltaic devices with uniquely designed nanostructures. In the artificial design systems, dye-sensitized solar cell systems were a base for investigating the above fundamental problems. For the efficiency of solar energy harnessing and better electron transport systems, uniquely assembled nanostructures of ZnO , TiO_2 , and an acceptor/donor energy system were fabricated using electrochemical approaches. The ZnO nanostructures including nanorods, nanosheets and nanowries were then decorated with small band gap semiconductors including, Fe_2O_3, CuO and Cu_2O by spin coating techniques. These structures were then investigated for electron transfer using characterization methods such as: SEM, TEM, XRD, and electrochemical potentiostat methods. Spectroscopic methods such as: UV-vis spectrometer, and time-resolved photoluminscence spectroscopy are being investigated now. This fundamental study will lead us to a better design for artificial photosystems.

470640 Direct Glycerol Fuel Cell with Polytetrafluoroethylene (PTFE) Thin Film Separator

Abstract:

Anion-exchange membrane direct glycerol fuel cells can yield a high power density, but challenges exist in developing chemically stable ion exchange membranes. In this work, we demonstrate a porous PTFE thin film, a well-known chemical, electro-chemical, and thermal robust material can work as a separator, and achieved high performance of direct glycerol fuel cells (DGFCs). DGFC using PTFE thin film with PdAg/CNT anode catalyst exhibit a remarkable peak power density of 214.7 mW cm^{-2}, which is about 22.6% lower than state-of-art AEM fuel cell with A901 membrane. The durability test was performed in order to evaluate the degradation rate of cell voltage of both PTFE thin film and A901 membrane. We report a 5.81% and 11.08% decrease in cell voltage for PTFE thin film and A901 anion-exchange membrane. Transmission electron microscopy (TEM) was used to evaluate the particle size of the anode catalyst after long-term discharge, and the results suggest that the PdAg particle size growth may be the main reason to lead performance drop. Our work shows, development of PTFE thin films could potentially alleviate the expense associated with anion exchange membranes for wide-spread fuel cell applications.

468826 Catalytic Behavior in Internal Steam Reforming of Methane in SOFC Conditions over $La_XSr_{2-X}MnO_4$ Materials

Abstract:

CATALYTIC BEHAVIOR IN INTERNAL STEAM REFORMING OF METHANE IN SOFC CONDITIONS OVER $La_XSr_{2-X}MnO_4$ MATERIALS

Karen Silva[1], Sebastian Vecino[1*], Monica Sandoval[1], Patrick Gèlin[2], Paola Gauthier-Maradei[1], Gilles Gauthier[1]

1: INTERFASE, Escuela de Ingeniería Química, Universidad Industrial de Santander, Carrera 27, Calle 9 Ciudadela Universitaria, Bucaramanga, COLOMBIA

2: Université Lyon 1, CNRS, UMR 5256, IRCELYON, Institut de recherches sur la catalyse et l'environnement de Lyon, 2 avenue Albert Einstein, F-69626 Villeurbanne, France.

*e-mail: jusevema13@hotmail.com

ABSTRACT

Nowadays, solid oxide fuel cells (SOFC) are one of the most attracting and interesting energy conversion devices. They can convert the chemical energy directly to electrical energy with high efficiency (>60%) and low environmental pollution, employing several types of fuels such as hydrogen, coal gas, natural gas (methane) or another hydrocarbons fuels. These electrochemical systems operate at high temperatures (600-1000°C) using oxygen from the air as oxidant, and hydrogen directly or by internal steam reforming (ISR) of some compounds present in the fuels to produce heat as well as water and electricity. The ISR is a highly endothermic process that takes place, catalytically, at the anode of the

SOFC. However, during this process and employing state of art anode material Ni/YSZ, the hydrocarbons decompose into hydrogen and additionally in an undesired compound: solid carbon (coke). This coke tends to deposit on the anode surface causing a rapid deterioration of the cell due to the deactivation of catalytic activity and the inhibition of hydrogen diffusion into the anode to continue with the electrochemical reaction. In order to overcome this limitation, it is necessary to develop new anode materials that catalyse the ISR and reduce significantly the problem described above. According to literature, single-phase oxide materials have been studied with excellent results. In this work, La0.5Sr1.5MnO4 was studied as potential new anodic material for SOFC directly fed with methane, operating at high temperature and under water deficient conditions. The material was synthesized by Pechini method and characterized by XRD. Measurements of catalytic activity in steam reforming of methane were carried out using 40 mg of catalyst at 900°C during 40h in a U-shaped quartz reactor. Different methane/water ratios (0.66≤CH4/H2O≤10) were employed to determine kinetic parameters. Coke formation during catalytic testing was measured by temperature programmed oxidation (TPO). The catalytic activity was compared with a reference material CGO (Ce0.9Gd0.1O2-d). Finally, the material was characterized again by XRD to evidence any damage or changes in the catalyst structure. This catalyst was found very active (high hydrogen production 9000ppmv) and resistant to coke formation (<69μmol/gcat), close to CGO catalyst (13000 ppmv hydrogen and 50.2μmol/gcat). The materials present an excellent stability during the test run; no structural change was found after the reaction. The small increase in the lattice parameters was attributed to the manganese reduction during the catalytic reaction (Mn+4/Mn+3 to Mn+3/Mn+2).

Key Words: *Solid oxide fuel cell, Internal steam reforming, Methane, Manganite, Energy.*

469639 Direct Fast Pyrolysis Bio-Oil Fuel Cell

Abstract:

Bio-oil derived from the pyrolysis of lignocellulosic biomass shows a great promise, however, needs further upgrading to potentially serve as an alternative to fossil fuels. Herein, we demonstrate that crude fast pyrolysis bio-oil can be directly used as a fuel for anion exchange membrane fuel cells (AEMFCs) to generate high power density electrical energy at low temperature. Direct fast pyrolysis bio-oil AEMFCs with the Pd/CNT anode catalyst and a commercial Fe-based cathode catalyst exhibit a remarkable peak power density of 42.7 mW cm^{-2}. Levoglucosan was identified as the major sugar compound with 11.1 wt% of the bio-oil composition, along with disaccharides, pyrolytic lignin, and oligomer of lignin-derived phenolic compounds. Cyclic voltammetry (CV) studies investigated the electrocatalytic oxidation of high purity levoglucosan over the four noble metal catalysts in half cell, as levoglucosan is the dominant sugar component in bio-oil. Pd/CNT, compared to other catalysts, displayed the highest activity and lowest onset potential of electrocatalytic oxidation of levoglucosan. AEMFC with high purity sugars shows ~ 1.2 to 3 times higher power density than that with fast pyrolysis bio-oil fuel.

461692 Nature-Inspired Fractal Flow Field for PEM Fuel Cells

Abstract:

Nature-Inspired Fractal Flow Field for PEM Fuel Cells

Jason In Sung Cho[1,2], Toby Neville[1,2], Jeffrey Marquis[3], Panagiotis Trogadas[1], Dan Brett[1,2] and Marc-Olivier Coppens[1]

[1] *Department of Chemical Engineering, University College London, London WC1E 7JE, United Kingdom, m.coppens@ucl.ac.uk*

[2] *Electrochemical Innovation Lab, Department of Chemical Engineering, University College London, London WC1E 7JE, United Kingdom*

[3] *Isermann Department of Chemical & Biological Engineering, Rensselaer Polytechnic Institute, 110 8th St., Troy New York, United States*

One of the ongoing challenges faced by polymer electrolyte membrane (PEM) fuel cells for broader commercialisation is the accomplishment of uniform gas distribution across the catalyst layer. Ensuring reactant homogeneity is essential to improve fuel cell efficiency and to forestall adverse side reactions that are detrimental to fuel cell longevity [1]. Within the confines of the conventional flow fields, such as serpentine channels, this is an unavoidable consequence, due to reactant depletion across the channel and over the active area as a whole [2].

This work aims to address the issue by implementing a novel 3D flow field inspired by the fractal geometry of the upper respiratory tract of the human lung. The human lung has been shown to have an optimal architecture that uniformly distributes oxygen from the trachea to the alveoli. Furthermore, the human lung transitions between two flow regimes: 14-16 upper generations of branches dominated by convection, and 7-9 lower generations of space-filling acini dominated by diffusion[3, 4]. The upper generations of branches are designed to slow gas flow to a rate compatible with the rate in the diffusional regime (Pe≈1) [5] and the entropy production was shown to be uniformly distributed in both regimes [3, 4]. Furthermore, an attractive aspect is the scalability of fractal networks. They can bridge multiple length scales by the addition of further generations, while preserving the building units and microscopic function of the system [6]. These characteristics that make the lung an optimal transport structure are all desirable for a PEM fuel cell. However, rather than uniformly distributing oxygen in a given volume, we wish to distribute the gases uniformly over a plane to ultimately deliver a uniform concentration of reactant throughout the catalyst layer [7].

The fractal pattern consists of repeating "H" shapes and, by employing a fractal dimension of 2, where daughter "H's" are half the parent's size, a plane-filling geometry is attained. Furthermore, the three-dimensional branching structure allows only the outlets of the fractal inlet channel to be exposed to the MEA, eliminating the reactant depletion along the channel, and thus providing uniform local conditions on the surface of the catalyst layer.

In this study, the reactant flow and cell performance were modelled for different numbers of fractal branching generation. Optimal performance was reached for branching levels N ³ 5. Based on the simulation results, three fractal flow field prototypes were produced via selective laser sintering (SLS), a «3D printing» method, each representing a branching generation level of N = 3, 4 and 5, and validated against a standard serpentine flow field.

The results suggest improved mass transport for the fractal flow fields at high current density, due to the homogeneous gas distribution. The fractal prototype with N = 4 generations demonstrated improved performance over the serpentine at all operating conditions investigated in this study and a good correlation with the results predicted by the model. The worse-than-expected performance of for N = 5 generations at higher humidity level reveals insufficient convective liquid water removal due to the slow air velocity from each fractal outlet hampering effective diffusion within the porous medium. The issues with flooding were alleviated by operating the fuel cell at lower humidity levels, which led to improved mass transport for the prototype with N = 5 generations. However, fuel cells should ideally be operated under fully humidified condition to minimise local membrane drying and subsequent Ohmic losses from reduced proton transport.

The findings elucidate the need for implementation of active water management systems to alleviate flooding for the fractal flow fields, especially at higher generation levels where flooding is projected to be more pronounced.

References

[1] Liu, Z., Yang, L., Mao, Z., Zhuge, W., Zhang, Y., and Wang, L., *Behavior of PEMFC in starvation*. Journal of Power Sources, **2006**. 157(1): p. 166-176.

[2] Tüber, K., Oedegaard, A., Hermann, M., and Hebling, C., *Investigation of fractal flow-fields in portable proton exchange membrane and direct methanol fuel cells*. Journal of Power Sources, **2004**. 131(1–2): p. 175-181.

[3] Kjelstrup, S., Coppens, M.-O., Pharoah, J., and Pfeifer, P., *Nature-inspired energy-and material-efficient design of a polymer electrolyte membrane fuel cell*. Energy & Fuels, **2010**. 24(9): p. 5097-5108.

[4] Trogadas, P., Ramani, V., Strasser, P., Fuller, T.F., and Coppens, M.-O., *Hierarchically Structured Nanomaterials for Electrochemical Energy Conversion*. Angewandte Chemie International Edition, **2016**. 55(1): p. 122-148.

[5] Coppens, M.-O., *A nature-inspired approach to reactor and catalysis engineering*. Current Opinion in Chemical Engineering, **2012**. 1(3): p. 281-289.

[6] Trogadas, P., Nigra, M.M., and Coppens, M.-O., *Nature-inspired optimization of hierarchical porous media for catalytic and separation processes.* New Journal of Chemistry, **2016**.

[7] Marquis, J., Nature-inspired hierarchically structured high-efficiency PEM fuel cell, PhD Thesis, **2013**.Rensselaer Polytechnic Institute, Troy, NY.

Acknowledgements

The authors wish to thank the National Science Foundation Fuel Cell IGERT at Rensselaer, DGE-0504361 for supporting initial work. Financial support from an EPSRC "Frontier Engineering" Award (EP/K038656/1) and a UCL Faculty of Engineering Sciences Dean's Scholarship are also gratefully acknowledged.

466171 Membraneless Fuel Cells Modified with Gold Nanoparticles and Laccase for Direct Conversion of Recalcitrant Carbohydrates into Electricity

Abstract:

The need for clean and viable sources of alternative energies has prompted the development of biofuel cells (BFCs), which are characterized by renewability and high power density. In particular, enzymatic BFCs are a special type of fuel cell in which enzymes are used as catalysts to oxidize specific substrates on anodes and reduce O_2 on cathodes. Candidates for substrates include simple sugar like glucose to complex ones such as starch and cellulosic biomass. Since most readily available biomass is in the raw, unprocessed cellulosic form, there is a challenge for BFCs to utilize these complex and recalcitrant substrates. Furthermore, more complex feedstocks such as hydrolysates from lignocellulosic biomass, however, have rarely been used as substrates in BFCs. Typically, some pretreatment or saccharifications are required for these complex feedstocks before they can be used as substrates to generate a high current density. Thus, there is motivation to design, construct and modify BFCs such that they are able to efficiently and effectively utilize various carbohydrate substrates.

In this study, we aim to address problems of using complex substrates by leveraging the catalytic properties of gold nanoparticles (AuNPs) on cathodes and laccase on anodes in a membraneless hybrid BFC. AuNPs are versatile in catalyzing a wide variety of substrates and present a viable alternative to substrate-specific enzyme catalysts that are traditionally used. This is beneficial especially when mixed substrates derived from biomass are used as fuel. We first fabricated and modified cathodes with AuNPs and characterized the effect of the presence of AuNPs (as compared to unmodified electrodes and that modified with bulk gold), AuNP sizes and loadings on the catalytic action of the BFC. Next, the harnessing of stable and useful currents from a variety of carbohydrate substrates was studied: In addition to glucose, other carbohydrates including recalcitrant disaccharides and polysaccharides such as fructose, sucrose and carboxymethylcellulose were also found to successfully generate significant current densities. To further demonstrate the versatility of the BFC, complex feedstocks substrates such as sugarcane bagasse hydrolysate was also shown to deliver a stable current, which was further validated with the lighting up of devices such as light emitting diodes.

476420 Role of Hydrogen in Future Clean Energy Landscape

Abstract:

Given the future energy demand growth, sustainable oil constraints and global mandates against GHG induced climate change; hydrogen **is** poised to be a strong contender in the future energy landscape. Strategic scenarios and **hydrogen roadmaps** have been laid out for addressing the future energy needs against the backdrop of environmental targets and clean energy strife. Hydrogen's appreciable leverage and share is put on the transport and power sector with hydrogen pathways rendering a strong impetus in the longer run.

Hydrogen's role as a unique carrier of ultra clean energy is imminent despite any barriers or roadblocks put by skeptics debating its claimed viability. Based on its unparalleled leverage and source flexibility, whether fossil, renewable or nuclear into 'one clean energy platform', the **hydrogen economy** is destined to be in the offing.

Extensive deliberations for transitioning from direct hydrogen substitution as fuel (like H-CNG or HICE applications) and **fuel-cells** based stationary distributed power to transport vehicles starting with fleets and marine applications, is now expanding to a sizeable deployment for personal (HFC) cars. Marking the onset of the hydrogen movement, Toyota had recently launched their personal HFC car "Mirai" last year in California, thus projecting an exponential expansion in coming years.

Furthermore, in view of the setback and limitations on nuclear power and also renewable energy options in terms of large scale deployment and overall economics, potential future trajectories are being forecasted on **hydrogen** as a dominant energy carrier in the longer term.

For realization of such demand, the fossil-derived hydrogen is still the strong contender and anticipated to cater for a substantial portion of future hydrogen generation, especially large centralized facilities and several medium-sized units, with carbon capture provisions.

It has been further boosted given the drive to exploit the abundant low-cost unconventional gas as a cost-effective low-carbon feedstock for generating large scale hydrogen using state-of-the-art technology. Additional drivers are ultimate well-to-wheel efficiency, economy of scale and qualification for CO_2 capture and sequestration. Eventual options of using renewable power to generate "**green hydrogen**" are also on the horizon.

The talk will provide an overview on these interesting facets on the role of hydrogen in the future energy landscape.

450795 Realization of Sustainable and Cost-Effective Cellulosic Ethanol Production By Deploying Enzyme and Advanced Yeast Technologies in an Integrated Approach

Abstract:

Ethanol is considered the most potential next generation automotive fuel because it is carbon-neutral and could be produced from renewable resources like lignocellulosic biomass. Key to economic viability of cellulosic ethanol is cheap production and application of different enzymes hydrolyzing lignocellulosic carbohydrates into fermentable sugars and efficient conversion of the derived monomeric sugars into ethanol at high productivities and yields using advanced yeast strains. DSM is a frontrunner in the development of bioconversion and fermentation technologies for the production of cellulosic ethanol, enabling the industry to diversify from starch crops to lignocellulosic agricultural residues, unlocking the full industrial potential of this cheap, abundantly available and sustainable type of feedstock.

Within DSM, research projects aim at developing both commercially competitive cellulolytic enzymes for the liquefaction and saccharification processes, and genetically engineered yeast strains that ferment hexoses and pentoses to ethanol. DSM's enzyme cocktail is uniquely active at high temperature and low pH, offering many benefits and product differentiation. Further strain and process development, enzyme cocktail ratio optimization, protein and genetic engineering have led to significant improvements in productivity, efficiency and costs. Fermentation of hexoses and pentoses from lignocellulosic hydrolysates by DSM's yeast strains has been significantly improved by applying both rational and evolutionary engineering strategies, resulting in C6- and C5-sugars fermenting *Saccharomyces cerevisiae* strains being commercially available. Additionally, next generation advanced yeast strains, which have accelerated pentose utilization, improved robustness, as well as the ability to convert inhibitors and specific biorefinery waste products into additional ethanol, are currently finding their way to commercial implementation.

In this context, it is of great importance to fully integrate the design, construction and operation of the aforementioned technologies into a biorefinery for biofuels production. In this presentation, we will discuss the latest developments and how those fit within the platform of on-site manufacturing of enzymes and yeast fermentation.

468107 Alternative Transport Fuels and Their Production Using Surplus Renewable Electricity, Water and CO_2

Abstract:

Electrical power generated from wind energy fluctuates strongly. As put forward by the EU Commission, in 2050 80 % of the electricity demand should be covered by renewables. Therefore, the integration of fluctuating electricity generation from renewables plays a fundamental role in Europe's and especially Germany's future energy system. Additionally for 2050, the EU Commission postulates a reduction of anthropogenic CO_2 emissions of about 80-95 %, compared to 1990. In Germany, the transport sector was accountable for approximately 20 % of the domestic CO_2 emissions in 2014, while contributing 30 % to the domestic energy demand. One reason for this high share is that 90 % of the transport sector is run on fossil fuels. Hence, to achieve the goals for 2050, non-fossil alternative fuels need to be found. A popular option is the further integration of fuel cell cars. However, especially for heavy duty vehicles and planes, there is long term need for liquid fuels such as diesel and kerosene. Another promising option is the idea of Power-to-Fuel (PTF): the synthesis of fuels based on sustainably produced hydrogen and usage of CO_2 as a feedstock. The energy required to run those processes can be provided by surplus electricity from renewable sources. This is energy which cannot be fed into the grid when there is no demand on the consumer side.

If an alternative renewable fuel is able to penetrate the market, two main requirements need to be met: its potential for an implementation strategy into an existing market as well as a technically feasible and economic production. Put differently, a renewable fuel should not require a significant alteration of the fuel infrastructure, thereby allowing a gradual integration in the existing infrastructure. Additionally, a range of different fuels for different usages need to be offered. Just as the fossil fuel market, the renewable fuel market needs diversification.

At the Institute of Energy and Climate Research (IEK-3) of Forschungszentrum Jülich, intensive research is being carried out on a comparable overview of the several products and process routes of PTF. In this contribution, investigations on two different routes from synthesis gas to an alternative renewable fuel will be outlined. The respective products of the two outlined processes are Fischer-Tropsch (FT)-diesel and polyoxymethylene dimethylene (POMDME). Both products can be blended with conventional diesel and with each other because they have appropriate compatibility. Therefore, an implementation strategy into the existing fuel market is practicable.

Despite an equal product application, the comparison of both technologies is widely diversified. At the molecular level, FT-diesel is closely related to conventional diesel; this applies less to POMDME. POMDME also has a considerably lower heating value; however, it has great potential as a fuel due to its beneficial combustion properties.

Although nowadays FT is not used for PTF in large scale, state-of-the-art FT-processes have been used industrially for many decades. In contrast to FT, owing to low product demand, processes for POMDME have not yet been developed on an industrial scale. Both processes have various options for the interconnection of system components and for complete process control. The system's interconnections as well as the process parameters have influence on raw material and energy demand and also product quality, co- and byproducts. Just as the product specifications, the process analysis results will play an important role in the fuel implementation strategy.

462845 Co-Products from the Catalytic Fast Pyrolysis of Biomass

Abstract:

Catalytic fast pyrolysis (CFP) of biomass is a potentially viable process for producing drop-in, renewable hydrocarbon transportation fuels and the economics may be improved by providing alternative pathways to valuable co-products. This provides flexibility to a biorefinery to adjust conditions to improve economics or to valorize waste low value streams. As an example, CFP using zeolite catalysts such as HZSM-5 produces aromatic molecules with 1-4 rings. The single ring products are excellent additives for gasoline because of their high octane numbers and their high heats of combustion. However, the larger aromatics have boiling points too high for gasoline and low cetane numbers, eliminating them as potential diesel additives. Hydrotreating these molecules is required to convert them to a fuel, but they may have value as co-products. Likewise oxygenated intermediates and products, such as phenols, formed during CFP may be high value products that can also help the economics of hydrocarbon fuels. These oxygen functionalities are often difficult to add to hydrocarbon molecules extracted from petroleum and producing these molecules from biomass may be intrinsically more environmentally benign. In addition, leaving oxygen in the molecules may improve the yields of biofuels and bioproducts, which would again improve the economics. We will discuss studies to improve selectivity of co-products and efforts to separate them and prepare polymers.
Catalytic fast pyrolysis (CFP) of biomass is a potentially viable process for producing drop-in, renewable hydrocarbon transportation fuels and the economics may be improved by providing alternative pathways to valuable co-products. This provides flexibility to a biorefinery to adjust conditions to improve economics or to valorize waste low value streams. As an example, CFP using zeolite catalysts such as HZSM-5 produces aromatic molecules with 1-4 rings. The single ring products are excellent additives for gasoline because of their high octane numbers and their high heats of combustion. However, the larger aromatics have boiling points too high for gasoline and low cetane numbers, eliminating them as potential diesel additives. Hydrotreating these molecules is required to convert them to a fuel, but they may have value as co-products. Likewise oxygenated intermediates and products, such as phenols, formed during CFP may be high value products that can also help the economics of hydrocarbon fuels. These oxygen functionalities are often difficult to add to hydrocarbon molecules extracted from petroleum and producing these molecules from biomass may be intrinsically more environmentally benign. In addition, leaving oxygen in the molecules may improve the yields of biofuels and bioproducts, which would again improve the economics. We will discuss studies to improve selectivity of co-products and efforts to separate them and prepare polymers.

480817 Application of Molecular Characterization to Bituminous Crude Oil to Study Asphaltene Precipitation

Abstract:

Application of molecular characterization to bituminous crude oil to study asphaltene precipitation

Md Rashedul Islam, Yifan Hao, Meng Wang, Toni Kirkes, and Chau-Chyun Chen

Department of Chemical Engineering, Texas Tech University

Abstract

Bituminous crude oils are vulnerable to asphaltene precipitation when blended with paraffinic samples. Precipitation behavior depends on the nature and content of the paraffins in the blend. A thermodynamic modeling of asphaltene precipitation from crude oil is important as part of the larger effort to understand the compatibility of bitumen or heavy crude oils with the paraffinic ones. Recently, asphaltene precipitations are predicted from wide spectrum of solvents by aggregation thermodynamic framework [1] developed from Yen-Mullins model [2]. However, precipitation prediction from crude oils is still very difficult because of lack of reliable techniques to represent petroleum fluids with real molecules. In this work, a molecule based characterization approach [3] is utilized to represents Cold Lake bitumen with real hydrocarbon molecules. Additionally, asphaltene precipitations are predicted form mixing n-alkanes with Cold Lake bitumen.

References

[1] M. Wang, Y. Hao, M. R. Islam, and C.-C. Chen, "Aggregation thermodynamics for asphaltene precipitation," *AIChE Journal,* vol. 62, pp. 1254-1264, 2016.

[2] O. C. Mullins, "The Modified Yen Model†," *Energy & Fuels,* vol. 24, pp. 2179-2207, 2010.

[3] C.-C. Chen and H. Que, "Method of Characterizing Chemical Composition Of Crude Oil For Petroleum Processing," US Patent Application Number: 2013/0185044, July 18,2013.

464593 Rapid Analysis of Syngas Using Micro GC Fusion

Abstract:

Rapid Analysis of Syngas using Micro GC Fusion

In 2012, the World Bank estimated that 140 billion cubic meters (bcm) of the 3370 bcm of natural gas produced worldwide was flared. Flaring burns surplus gas and is detrimental to the environment, contributing over 350 million tons of CO2 in annual emissions. An alternative to flaring is to convert the associated natural gas into synthesis gas (syngas), which is an intermediate gas used to produce synthetic liquid fuel. Converting syngas to liquid chemicals through the Fischer-Tropsch (FT) process presents an economical and greener alternative to flaring. Syngas is composed predominantly of hydrogen and carbon monoxide, though it may also contain carbon dioxide, nitrogen, methane, ethane, ethylene, or H2S in the high ppm to percent level. After catalytic conversion, determining the syngas composition is a critical step before using the gas in any downstream process. To ensure efficiency, the analysis of syngas must be analyzed quickly and accurately. Building on proven microelectromechanical systems (MEMS) technology, Micro GC Fusion is capable of providing rapid analysis of typical syngas components in a compact instrument. A method equipped with temperature programming provides an area %RSD of less than 1.0% repeatability. This accurate and repeatable analysis of syngas is performed in less than 50 seconds.

468268 Molecular-Level Kinetic Model Informed Life Cycle Analysis of Green Diesel Production

Abstract:

The drive to reduce our reliance on petrodiesel coupled with the advent of new environmental policies and economics has prompted the production of green diesel as an alternative, sustainable fuel. In this study, a molecular-level kinetic model was used to investigate the impact of changing feedstock composition and reaction conditions on the life cycle analysis of green diesel. The kinetic model simulates a fixed bed hydrotreating unit that includes the detailed triglyceride and free fatty acid hydroprocessing chemistry to produce paraffin-rich diesel range fuels. Linear free-energy relationships and literature data were used to reduce and tune the parameters of the LHHW rate laws allowing for product cloud point and cetane number property calculations. The product green diesel has excellent combustion properties but poor cold-flow properties, so hydrocracking severity was used to regulate end use and value of the green diesel. The process inputs and outputs were used in Argonne National Laboratoryâ€™s Greenhouse Gases, Regulated Emissions, and Energy Use in Transportation (GREET) model to evaluate greenhouse gas emissions of green diesel compared to petrodiesel. Scenario testing was performed to study the relative emission reduction for the different reaction conditions.

470164 Application of the Advanced Distillation Curve Method to Characeterize Two Alternative Transportation Fuels Prepared from the Pyrolysis of Waste Plastic

Abstract:

The properties of alternative fuels must be evaluated to determine their suitability for use in existing engines, either as blends with conventional petroleum products or as stand-alone substitutions (that is, drop in replacements). Alternative fuels made from waste feedstocks are especially attractive since they address the dual purpose of fuel production and mitigation of environmental pollution environmental concerns. In this study, two alternative liquid fuels generated by pyrolysis of waste polyethylene plastic bags were assessed using a composition-explicit distillation approach, which provides a rich data matrix that includes including fuel volatility, composition, hydrocarbon classification, and energy content in each distillate volume fraction. The data collected using this approach provide valuable information for the development of equations of state to describe the thermodynamic behavior of these complex fluids. The analysis showed that, despite significant differences in composition, the distillation curves and associated data of the alternative fuels show promising similarity to their conventional counterparts.

469215 Application of Real-Component Based Molecular Characterization on Petroleum Refinery Simulation

Abstract:

Abstract

As light to medium sweet crude oils keep depleting while demand of oil products are increasing, heavier and sourer crude has become a growing component to global oil market. Accompany with that, extreme bulk properties (including high molecular weights, high boiling points, large densities, high viscosities, etc.) are expected. However, current experimental approaches are still limited in characterizing heavy crudes in two aspects: 1) chemical structures of hydrocarbons with carbon number up to merely 12 can be accurately analyzed; 2) distillation cuts can only be obtained up to 750 K of normal boiling point, even with deep vacuum fractionation apparatus[1]. With limited experimental data available, less than 50 wt. % of individual heavy oil or bitumen can be analyzed, which renders simple model extrapolation over the whole crude highly unreliable.

The present work compares performance of petroleum refining simulation between two different methods: 1) the traditional API correlations commonly used in industry, 2) a real-component based molecular characterization method[2]derived from thermodynamically-consistent correlation of available crude oil data on chemical and physical properties. Compared with the former method, the later one not only gives reasonable predictions for properties of crude oil including heavy cuts, but also provides chemical structures and compositions of model constituent components over the whole crude oil, which serves as a cornerstone of high fidelity refining simulation.

Reference

1. Sánchez-Lemus, M. C., et al. (2016). "Physical properties of heavy oil distillation cuts." Fuel 180: 457-472.

2. Chen, C.-C. and H. Que (2013). "Method of characterizing chemical composition of crude oil for petroleum refinery processing". US Patent Application 20130185044 A1.

460128 Economic Design of Bioconversion Process: Natural Gas to Isobutanol

Abstract:

The low cost and abundant supply of natural gas have been motivators for the development of many technologies that convert this low cost carbon feedstock into compounds of greater value. Natural gas bioconversion is a developmental technology which applies synthetic biology to a naturally methane consuming bacteria, a methanotroph, to program it to produce higher value materials of interest. Intrexon has formed joint ventures to commercialize methanotrophic bioprocesses for biofuels and high value chemical intermediates.

Using a methanotrophic bioprocess to produce fuels and chemicals has several advantages other than the low cost and minimal processing of feedstock. Natural gas bioconversion offers scaling opportunities beyond traditional gas-to-liquid (GTL) facilities, as well as biomass feedstock facilities. Theoretical yields of methanotroph/methane exceed that of yeast/sugar approaches by a wide margin. Biocatalytic processes that achieve high yields under milder processing conditions allow for reduced capital and operation expenditures, offering greater economic viability.

Intrexon's flagship molecule of interest in the methanotroph platform is isobutanol, a fuel for gasoline blending which has several positive attributes. Isobutanol is a clean burning "drop-in" fuel which can be blended with gasoline and is compatible with the existing petroleum infrastructure. Currently, provisions allow for bio-butanol blending up to 12.5% with gasoline and ASTM specifications exist for sale as fuel. Combustion of isobutanol has cleaner emissions as compared to gasoline by reducing hydrocarbon, carbon monoxide, nitrous oxide and sulfur dioxide production. Isobutanol also has advantages over first generation biofuels, such as ethanol, because of its higher energy density.

Currently, the production of both isobutanol and farnesene has been demonstrated in lab scale fermentation. To successfully bring this technology to commercial scale, consideration must be given not only to optimizing the metabolic pathway engineered into the methanotroph, but to key issues in the fermentor design, downstream processing and overall plant economics. Traditional engineering workflows, such as a front-end loading strategy, were used to shape the commercial scale process design. The biologic, technical and economic commercial feasibilities of the end-to-end natural gas to isobutanol process were critical to Intrexon's plant design. Laboratory scale learnings were applied to select technology that was commercially feasible at demonstration scale. This demonstration scale design was scaled down to pilot scale, where the operation will yield learnings into refinement of at-scale design, equipment selection, empirical data on heating and cooling, and controls strategies.

463172 Techno-Economic Analysis of Indirect, Direct, and Hybrid Coal-Biomass to Liquids (CBTL) Plants with CO_2 Capture and Storage (CCS)

Abstract:

Indirect coal liquefaction (ICL) and direct coal liquefaction (DCL) technologies are commercially available for producing alternative transportation fuels. However, both processes are plagued with high CO_2 emission. Because the syncrude produced in the ICL and DCL reactors cannot satisfy the current specification of transportation fuels, additional upgrading technologies are required. For example, syncrude from the ICL process has low octane number for naphtha cut but high cetane number for diesel cut, while syncrude from the DCL process has high octane number for naphtha cut but low cetane number for diesel. Therefore the hybrid coal liquefaction (HCL), a combination of ICL and DCL technologies, can reduce the penalty of downstream syncrude upgrading unit by optimal blending. Addition of biomass and application of carbon capture and storage (CCS) technologies are two possible solutions to reduce the carbon footprint, but would lead to higher operating cost and capital investment. Several studies have been conducted for indirect coal-biomass to liquids (CBTL) plants with CCS, while few studies have been conducted for direct and hybrid CBTL processes at the systems level. With this motivation, we will present a techno-economic study conducted in Aspen Process Economic Analyzer® (APEA®) environment for indirect, direct, and hybrid CBTL plants with CCS using high fidelity process models developed in Aspen Plus® and Excel.

In the indirect CBTL plant, coal and biomass are first gasified to syngas. Then the syngas is converted to syncrude via Fischer-Tropsch (FT) synthesis. CO_2 is captured from both raw syngas and FT vapor product. In the direct CBTL plant, coal and biomass are directly converted into syncrude in the catalytic two-stage liquefaction (CTSL) unit by adding hydrogen produced from gasification of coal/biomass/liquefaction residue or reforming of shale gas. Significant amount of CO_2 that is generated in the hydrogen production unit(s) is captured to satisfy the target extent of CO_2 capture. In the hybrid CBTL plant, pre-processed coal and biomass are sent to either syngas production unit or the CTSL unit. Produced syngas is sent either to FT unit or hydrogen production unit. Naphtha and diesel products from the FT unit and the CTSL unit are blended to reduce the syncrude upgrading penalty. Different CCS technologies are considered and optimized for the indirect, direct and hybrid CBTL plant depending on the sources of CO_2 containing stream and corresponding CO_2 partial pressure.

The focus of this presentation will be on optimal synthesis of the hybrid plant. In particular, following aspects will be presented: (1) development of plant-wide process model and rigorous economic model in Aspen Plus®, Aspen Exchanger Design and Rating®, Excel, and APEA®, (2) sensitivity studies to analyze the impact of key design parameters (i.e. biomass/coal ratio, extent of CCS, CCS technologies, blending ratio of indirect and direct syncrude in the hybrid route) and investment parameters (i.e. price of coal and biomass, project life, plant contingency and plant capacity) on key economic measures, such as net present value (NPV), internal rate of return (IRR) and break-even oil price (BEOP), (3) analysis of the potential environmental credits due to use of biomass and CCS for all liquefaction approaches, (4) comparisons and analysis of trade-offs of indirect, direct, and hybrid CBTL plants.

465496 Finished Fuel Blending Models for Assessing Integration of Biomass-Derived Products with Petroleum Refinery Products

Abstract:

As the bioenergy industry focuses on the near-term development of conversion pathways from biomass feedstocks to infrastructure-compatible fuel blendstocks, analysis capabilities and models are essential for assessing the values of biomass-derived blendstocks based on its fuel properties and the constraining parameters for blending them with petroleum-derived products. Funding from DOEâ€™s Bioenergy Technologies Office (BETO) has enabled the analysis team to develop both linear programming models (Aspen PIMS) and process models (Aspen Plus and Aspen HYSYS) to meet the new analysis needs. In addition to developing these analysis tools, the team has also utilized published experimental blending data to validate the results of the models and to improve blend property predictions. These models are already playing a vital role at NREL in understanding the potential to integrate fuel products from biomass conversion pathways under BETOâ€™s research and development portfolio.

This presentation will focus on (1) an overview of the analysis approaches and process models developed under this effort, (2) the benefits for utilizing the tools for assessing biomass-derived blendstocks, (3) examples of blending analysis results, (4) comparison of results from blending models relative to experimental data and (5) future work for this effort in assessing blending of oxygenated compounds.

474277 Catalytic Process for Levoglucosan Conversion into Distillate Range Molecules

Abstract:

Pyrolysis oils are low quality fuels which cannot be used in conventional gasoline or diesel engines largely due to their high oxygen contents. Upgrading strategies for deoxygenation of pyrolysis oils are imperative for use in transportation fuels. In this presentation, we describe a multi-step catalytic process for conversion of levoglucosan (LGA), a major anhydrosugar present in pyrolysis oil, to monofunctional compounds, which can subsequently be coupled to obtain distillate range molecules.

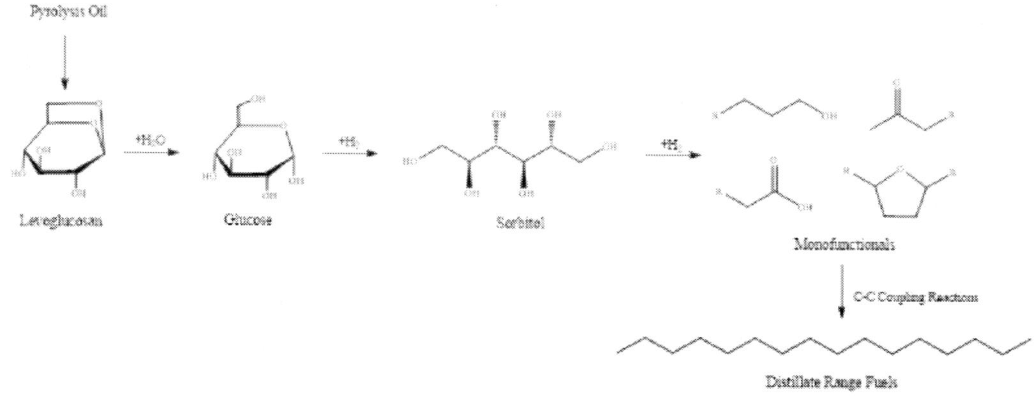

Figure 1. Pathway to convert levoglucosan to distillate range fuels

Sugar syrups produced via the pyrolysis of red oak pyrolysis oil are rich in LGA. Firstly, LGA in this sugar syrup is hydrolyzed to glucose. Following hydrolysis, glucose is converted to sorbitol with high yield by hydrogenation over Ru/C. Monofunctional intermediates are produced in the aqueous-phase hydrodeoxygenation (APHDO) of sorbitol with carbon yields up to 55%. These monofunctional species consist mainly of alcohols (71%) and heterocyclic species (24%). The majority of these species (52%) are C_5 or C_6 compounds, and only small amounts of ketones and acids are formed. The monofunctional intermediates are then converted into distillate range molecules by dehydration-oligomerization. The sorbitol APHDO effluent is fed to a flow-through reactor packed with a solid acid catalyst, SiO_2/Al_2O_3, to convert the monofunctional stream to olefins. Nearly 75% yield to mono-alkenes and dienes is obtained, based on the total carbon in the monofunctionals. Dienes originate from ring opening of heterocyclic compounds. C_2-C_6 olefins thus obtained can be oligomerized over acid catalysts to produce distillate range

hydrocarbons.

456831 Hydrodeoxygenation of Pinyon Juniper Catalytic Pyrolysis Oil to Hydrocarbon Fuels

Abstract:

As a renewable source, biomass is an essential option for diminishing dependence on conventional fossil fuel energy sources. Pyrolysis is a promising technology for the conversion of biomass into liquid fuels. However, several challenges associated with using pyrolysis oils such as their high acidity and low energy content inhibit their direct use as transportation fuels. We conducted a batch hydrodeoxygenation of pinyon juniper catalytic pyrolysis oil using Ni/SiO_2-Al_2O_3 catalyst to improve the following properties of the oil: heating value, acidity, oxygen content, water content, and viscosity. During the hydrogenation process, the influence of four experimental factors; temperature, catalyst loading, residence time, and hydrogen pressure was investigated. Once hydrogenation was completed, gas, coke, and a liquid product of two immiscible phases (aqueous and organic), were obtained. Maximum hydrogenation was obtained at a reaction temperature of 450 °C with catalyst loading of 20% (wt. % of total bio-oil feedstock), an initial hydrogen cold pressure of 1000 psi and a residence time of 30 minutes. At these conditions, bio-oil was deoxygenated by 96.17%. After hydrodeoxygenation, the higher heating value of the organic liquid product was 45.68 MJ/kg compared to 27.64 MJ/kg of the bio-oil. The water content of the organic liquid was zero compared to 1.63% of the bio-oil using Karl Fischer titration method. The aqueous fraction of the liquid product was 99.61% water. Furthermore, pH of the organic liquid was 6.87 compared to 3.46 of the starting material. The viscosity of bio oil was119.37 cP while it was 1.27 cP for the organic liquid product. The hydrogenation process provided a means of producing upgraded bio-oil, which possessed properties similar to that of gasoline.

458036 A Synthetic Biology Chassis for Lignin Valorization into Muconic Acid and Other Value-Added Chemicals

Abstract:

Lignin represents an untapped portion of biomass for the production of fuels and chemicals. In nature, many aerobic organisms degrade lignin-derived aromatics through conserved intermediates including protocatechuate and catechol. Employing this microbial approach offers a potential solution for the bioconversion of lignin into value-added chemicals for a future biorefinery. In this study, two hybrid biochemical routes combining chemical catalysis, lignin bioengineering, and metabolic pathway reconstruction were demonstrated for the conversion of lignin into value-added products. In the biochemical route 1, alkali lignin was chemically catalyzed for the release of vanillin and syringate as major products, which were further bio-converted into *cis, cis*-muconic acid and pyrogallol through engineered *E. coli* strains. The highest yields achieved were at 0.63 g ccMA/g vanillin in the fermentation broth and 0.69 g ccMA/g vanillin in whole cells mixture. While the highest yield of pyrogallol in the fermentation broth was 7.3 mg pyrogallol/g syringate. In the second biochemical route, the lignin biosynthesis pathway of plant *Arabidopsis* was engineered to accumulate protocatechuate (PCA) as an intermediate compound. The PCA extracted from the engineered *Arabidopsis* was further converted into ccMA using the engineered *E. coli* strains. In this biochemical route, the extraction yield of PCA was 1.45 mg PCA/ g plant biomass. The highest yield of ccMA was achieved at 0.31 g ccMA/ g PCA or 0.45 mg ccMA/ g plant biomass. In addition to ccMA, catechol was produced at a higher yield of up to 0.54 g catechol/ g PCA or 0.79 mg catechol/ g plant biomass. This study reports for the first time a direct process for converting lignin into ccMA and pyrogallol as value-added chemicals. It provides a great alternative pathway for the production of fuels and valuable from lignin.

468091 Porous Carbonaceous Solid Acids Derived from Farm Animal Waste and Their Use in Catalyzing Biomass Transformation

Abstract:

We demonstrate here the transformation of waste cow manure into N rich porous carbons (NPC) with large amount of micropores and large Brunauer–Emmett–Teller (BET) surface areas. The transformation process was achieved *via* carbonization of cow manure in presence of $ZnCl_2$ and $FeCl_3$ additives, without using additional templates. The resultant NPC could be converted into microporous carbonaceous solid acids (NPC-[C_4N][X]) by quaternary ammonization of samples with 1,4-butane sultone, and ion exchanging with strong acids such as HSO_3CF_3 and H_2SO_4. X-ray diffraction patterns and Raman spectra showed that NPC-[C_4N][X] exhibit amorphous carbon network. N_2isotherms and SEM images showed that NPC-[C_4N][X] have abundant nanopores with large BET surface areas. XPS, FT-IR and elemental analysis results showed the N rich network in NPC support, has been successfully and homogeneously functionalized with strongly acidic ionic liquid and sulfonic groups. ^{31}P solid NMR spectra confirmed that NPC-[C_4N][X] have typical Brønsted acid centers with strong and controllable acidity. Catalytic tests showed that NPC-[C_4N][X] exhibits excellent activity and reusability in production of biofuels and fine chemicals through esterifications, transesterifications, and depolymerization of microcrystalline cellulose. This approach may provide a novel and cost effective way to recycle the waste cow manure into efficiently porous carbon based solid acids for heterogeneous acid catalysis.

471686 Analysis of Carboxylic Acids in the Biocrude Derived from Hydrothermal Liquefaction of Lignocellulosic Biomass

Abstract:

The authors report (a) hydrothermal liquefaction (HTL) of lignocellulosic biomass using homogeneous nitrate salt catalyst: $Ni(NO_3)_2$, $Ca(NO_3)_2$, $Co(NO_3)_2$ and $Fe(NO_3)_3$ for the production of carboxylic acid enriched biocrude, and (b) detailed investigation of biocrude generated from a biomass emphasizing on the selectivity of lactic acid in particular and (c) study of the possible reaction pathways for production of carboxylic acids, biocrude and bio-oil with their energy content from a biomass. A biomass was loaded in the high temperature high pressure SS316 reactor and HTL of biomass was performed at 200 ^{o}- 300 ^{o}C. During HTL reaction, liquid samples were withdrawn at regular time intervals and the processed slurry was centrifuged to recover biocrude and residue. The biocrude was further analyzed by TOC analyzer, GC-MS and HPLC. The biocrude and residue obtained were further extracted in different solvents followed by evaporation on Rota evaporator to recover bio-oil and the bio-oil was analyzed using GC-MS. The HPLC analysis of biocrude revealed the presence of sugars, C_1-C_3 carboxylic acids such as lactic, propionic, acetic and formic, HMF/furfural and other oxygenated compounds. Among different C_1-C_3 acids, lactic acid was observed in the major amount of 83.92 wt% using pinewood: solvent of 1:75 at 250°C. The results obtained on carboxylic acid enriched biocrude, bio-oil composition, selectivity of organic acids such as lactic acid obtained during HTL of biomass substrates and possible reaction pathways will be presented in detail.

472041 Catalytic Pyrolysis of Sugar Cane Bagasse and Rice Straw

Abstract:

Proper utilization of agricultural wastes is particularly important not only for prevention of environmental pollution but also creation of value added by-products from them. In rural areas, the agricultural wastes is a good resource for animal feed if the transportation cost are low. If there is no local demand for such wastes burning in the fields is a common practice; which although is an easy solution for the farmers create air pollution for the close and distant localities. Sugar cane bagasse (SCB) and rice straw (RS) are potential environmental thread in countries where huge amount of such waste have no industrial utilization. Biochemical or thermochemical conversion technologies can be used to convert agricultural waste streams into biobased chemicals and biofuels. In this research, SCB and RS supplied from Egypt were pyrolyzed in a 2-inch fast pyrolysis reactor at 450°C. Commercial HZSM-5 and granulated red mud were used as the catalyst and nitrogen was used for fluidizing the system. The yields of SCB and RS pyrolysis liquids ranged from 35% and 28% over HZSM-5 catalyst and 36% and 24% over red mud catalyst respectively. The pH of the SCB and RS pyrolysis oils was 2.0-3.0 . The yields of char were 16% and 42% respectively for SCB and RS over HZSM-5 catalyst and 31% and 40% respectively for SCB and RS over red mud catalyst. The gas products were 30% (SCB) and 40% (RS) over HZSM-5 catalyst and 32% (SCB) and 36%(RS) over red mud catalyst. The majority of the gas were CO_2, CO and CH_4 together with H_2 and C_2-C_4 hydrocarbons at lower concentrations.

462072 Kinetics of Coffee Oil Ethanolysis Catalyzed By an Immobilized Lipase Enzyme to Produce Biodiesel

Abstract:

As supplies of fossil fuels are depleted, the need for alternative sources of fuel becomes evident. Biodiesel is an attractive alternative to petroleum-based fuel; however, cost of the feedstock, mainly vegetable oils, makes biodiesel a costly product. In the search for alternative feedstocks, spent coffee grounds were revealed to be a largely available and economical source of vegetable oil. To design a process for biodiesel production, a kinetic rate model was necessary. Immobilized lipase was found to be an optimal catalyst for the ethanol transesterification of coffee. Tests were performed in batches, a differential packed bed reactor, and a recycle reactor. Reactor effluents were analyzed by GC-MS. Results indicated significant external mass transfer limitations and stability of the immobilized enzyme catalyst at temperatures less than or equal to 45°C. A modified Michaelis-Menten equation for two substrates provided a good fit for the reaction kinetics in the range of 30° to 45°C.

473126 Different Parameters Controlling the Biocrude Yield in Hydrothermal Liquefaction of Microalgae

Abstract:

Research into alternate sources of fuel or renewable energy reveal many promising ideas for alternative energy sources. One of these possibilities is converting biomass to biofuel. Under high pressure and temperature, microalgae can be polymerized to biofuel. There are some thermo-chemical processes available to convert the biomass like pyrolysis and gasification, but these processes require dry feedstock. Thus these processes encounter large energy loss by vaporizing the moisture. Hydrothermal Liquefaction (HTL), a high temperature, high-pressure polymerization process, can convert wet biomass, including water to biocrude oil, biochar, and nutrient rich water. The biocrude obtained from HTL are environmentally friendly and can be directly used as fuel or can be upgraded to transport fuel. Many researchers are working on increasing the efficiency of HTL. In the present study an attempt is made to optimize the HTL reactor for maximizing the biocrude yield. Batch reactions are carried out to correlate the biocrude yield with different parameters like the reaction temperature (300Â°C - 350Â°C), reaction time (30 mins â€" 60 mins), initial solid concentration (5% - 20%) and also different species with different lipid contents. Experiments are also carried out to see the effect of different cultivating medium (wastewater and freshwater) on the crude yield after HTL. Experimental results show that different operating conditions can result in the biocrude yield to vary between 5% and 60%. These parameters are also related to the amount of heavier biocrude oil produced. This paper carefully examines and explains the relationships between the operating conditions and the biocrude yield to be able to optimize a HTL batch scale reactor for maximum crude yield.

451152 Synthesis and Characterization of Proto-Kerogen Via Simulated Diagenesis of *Litopenaeus Setiferus* and *Procambarus Clarkii* chitin

Abstract:

Petroleum formation begins with organic matter (OM). Chitin is a type of organic matter found in shellfish. Chitin is a long-chain polymer of an N-Acetyl Glucasamine, a derivative of glucose. Over geologic time and progressive burial, chitin and other OM undergo sequential catalytic modifications known as diagenesis, catagenesis, and metagenesis, resulting in the petroleum-forming intermediate known as kerogen. Catalytic activity is provided by aluminosilicate clays such as kaolin, feldspar, bentonite, and wollastonite. With progressive burial, temperatures eventually rise into the "oil window" of 100-200°C. Within the oil window, organic matter is gradually converted into complex intermediates including fulvic and humic acids, eventually forming kerogen.

For this experiment chitin is isolated from *Litopenaeus setiferus* and *Procambarus clarkii* by sequential treatment with 1M hydrochloric acid followed by 1M sodium hydroxide. Reactors are then mixed with various aluminosilicates and incubated in a temperature-controlled oven. Experimental conditions evaluated in this research include residence times ranging from 1 day to 1 month and reaction temperatures of 125°C, 155°C, and 175°C. The impact of pH on reaction products is also evaluated in this research. The products of the reaction are characterized using thermogravimetric analysis, which measures sample volatility with respect to temperature. Specific molecules produced in each reaction are then identified by NMR.

468959 The Impact of Treatment Conditions on Biomass Maturation and Conversion

Abstract:

Kerogen is the natural product of the maturation of organic matter. Over time and progressive burial, biomass undergoes sequential catalytic modifications known as diagenesis, catagenesis, and metagenesis. The aluminosilicate clays that co-deposit with organic matter provide catalytic function.

The types of biomass evaluated in this study include lignin, bagasse, yeast, chitin, and algae. Lignin was obtained by treating sugarcane bagasse with sulfuric acid to hydrolyze cellulose and hemicellulose. Chitin was isolated from crawfish and shrimp casings by sequential treatment with HCl and NaOH. *Chlorella Protothecoides* was grown phototrophically and *Saccharomyces cerevisiae* was grown heterotrophically. Biomass samples were then mixed with an aluminosilicate.

Biomass-aluminosilicate samples were loaded into isochoric reactors and incubated at elevated temperatures for the desired resident time. Different aspects of the reactions were changed to determine the various effects on pre-kerogen. The temperatures evaluated were 175°C, 150°C and 125°C. Aluminosilicate used were wollastonite, bentonite, and feldspar. Residence times ranged from 24 hours to 28 days.

The results of these variables will be presented with thermogravimetric analysis (TGA). TGA measures the volatility of a sample as temperature is increased. In this study, TGA was used to evaluate the impact of the treatment conditions outlined above on the organic products of simulated diagenesis. Characteristic volatility curves were used to determine the temperature ranges by which fractions of the biomass volatilize before and after maturation experiments.

471691 Ethanol Conversion to n-Butanol Via the Guerbet Chemistry: Role of Mixed Oxide Catalysts

Abstract:

In future ethanol is expected to be available in large quantities beyond the transportation blending needs. The excess ethanol can be converted over mixed oxide catalyst to butanol and higher alcohols via complex tandem reaction mechanism (Guerbet) which can be used as a chemical or fuel intermediate. The cascade nature of the Guerbet chemistry requires catalytic materials containing active sites with different functionalities at unique molecular distances. The hydrotalcite derived MgO-Al_2O_3 mixed oxide catalyst is identified as one of the efficient material to convert ethanol to n-butanol and higher alcohols. The catalyst preparation method, mixed oxide composition and the experimental conditions plays a role in directing the ethanol condensation chemistry. In this talk the conversion of ethanol to butanol over a multi-functional mixed oxide catalyst with respect to the catalyst synthesis and preparation, reaction conditions and the reaction mechanism will be discussed.

461389 Microwave-Assisted Pyrolysis of Coal-Biomass Mixturs UNDER Reactive Gases

Abstract:

Conversion of coal/biomass blends into higher value chemicals can either be achieved directly, by non-oxidative routes or indirectly, by oxidative routes. Usually, multiple chemical processes are involved in oxidative routes; first a gasification step is needed to produce syngas, followed by Fisher Tropsch reactions to produce chemicals. On the other hand, pyrolysis could be used as a single step process to convert these blends directly into chemicals. However, the main obstacle in conventional thermal pyrolysis is the formation of high molecular weight tars which is difficult to handle and require further cracking to be utilized. In this work, microwave-assisted pyrolysis was studied on samples of Mississippi lignite and PRB subB coals mixed with 25 wt% yellow pine wood sawdust. The microwave-assisted pyrolysis product distributions of the above blends were compared to those using conventional thermal pyrolysis. . Microwave energy was not only used for rapid heating but also to examine any possible kinetic enhancements of this process either under inert (N_2) or reactive hydrogen rich (H_2 and CH_4) gases. Our initial GC-MS results suggested that the oil product distributions were shifted to relatively lower molecular weight chemicals using microwave heating compared to conventional heating. It also suggested that the addition of hydrogenated gases increases the oil yield produced. The chars produced were characterized by different techniques, such as SEM and XRD to determine the morphology and crystallinity of these samples before and after pyrolysis, respectively. The porosity of fresh and char samples were determined using N_2-adsorption surface area measurements.

461602 Computational Modeling of Electrochemical Bio-Oil Upgrading

Abstract:

Electrochemical hydrogenation (ECH) reactions are needed in the post-pyrolysis treatment of bio-oil to convert biomass into biofuels. A computational approach is taken to find the optimal catalyst material, and the conditions at which the concentration of species to react surrounding the catalyst is best. This is done to provide testable hypotheses regarding catalyst activity and selectivity for experimental work. Despite extensive research done in electrochemical reactions, the complex solvent/catalyst environment has hindered a rigorous understanding of electrocatalyst surface chemistry. Density functional theory calculations of species binding and ECH reaction energies were done as a function of voltage on the surface of numerous catalysts. Initial results show that Au, Cu, and C catalysts are best suited because they preferably bind organic molecules over hydrogen, and show a large applied potential for H_2 formation and low one for organic hydrogenation. Since only the species at the cathode vicinity are susceptible to undergo electron transfer, classical molecular dynamics simulations of solvent mixtures in an electrolytic cell were performed to assess how species concentrations differ between the solid/liquid interface and bulk regions. Simulations were performed at varying temperatures, cathode surfaces and charge. Energetic and entropic contributions were analyzed to understand competing surface binding and solvation effects. Initial results show a decrease in organic and increase in water mol fractions at the solid/liquid interface as the cathode charge in the electrolytic cell increases. This can only be compensated by a strong adhesion of the organic molecule to the cathode surface. Results are compared with concurrent experiments and implications on catalyst choice are discussed.

472694 Optimizing Catalytic Fast Pyrolysis and Hydrotreating for the Production of Biofuels

Abstract:

Liquid transportation fuels can be produced via catalytic fast pyrolysis by upgrading biomass pyrolysis vapors prior to condensation. While it is possible to obtain a completely deoxygenated hydrocarbon product by using suitable operating conditions, including high biomass-to-catalyst mass ratios, the yield of the product oil remains low. Higher yields can be obtained at lower biomass-to-catalyst ratios but the resulting oil includes oxygenates and hydrotreating is required for the final deoxygenation. With decreasing degree of deoxygenation in the catalytic pyrolysis step, higher severity is required for hydrotreating. This research discusses the optimization of the overall carbon efficiency and cost of the process for producing oils with low oxygen contents via catalytic fast pyrolysis. Pine was pyrolyzed and upgraded over ZSM-5 catalyst in a dual 2-inch fluidized bed reactor system. The catalyst-to-biomass ratio was varied to produce oils with different oxygen contents, and the oils were subsequently hydrotreated over a sulfided CoMo catalyst.

470848 In Situ Raman Analysis for the Production of Biofuel Using a Heterogeneous Layered Double Hydroxide Catalyst

Abstract:

Oral Presentation

In Situ Raman Analysis for the Production of Biofuel Using a Heterogeneous Layered Double Hydroxide Catalyst

Obakore Agbroko , Keyvan Mollaeian, William E. Holmes, Tracy J Benson

As the world's energy consumption is rapidly increasing many efforts are being made to produce renewable and sustainable fuel sources that provide independence, or at least displacement, from fossil fuels. Many of these renewable sources include sunlight, wind energy, geothermal energy, and the conversion of plant based biomass into biofuels. The conversion of vegetable oils into biodiesel is a highly viable option for supplementing fossil fuel energy. Unfortunately, most biodiesel is produced from oils stemming from row-crops, which take up viable land resources needed for food crops. One promising alternative is microbially-produced oils (MPOs) where microorganisms feed on industrial and municipal wastewaters. MPOs, however, contain high concentrations of free fatty acids (FFAs), unlike most row-crops. Under conventional biodiesel processes, high concentrations of FFAs in oil feedstocks significantly complicate conversion and separation of the fuel produced. However, if dimethyl carbonate (DMC) is used as a substitute for methanol, the formation of an unwanted glycerol byproduct is avoided. In addition, the use of a heterogeneous catalyst, in our case triazabicyclodecene bound to Mg/Al layered double hydroxide (TBD-LDH), process wash water is not required for separation of fuel components from catalyst salts.

In this study, an in situ Raman analysis method was developed to monitor the reaction conversion as well as kinetic parameters for the transesterification of lipid oils to biofuel using DMC and TBD-LDH catalyst. This in situ method, using a fiber optic probe attached to the 785 nm Raman system, does not require sample handling and workup as needed for GC analysis. The height of the C=C stretching mode of unsaturated fatty acids of the oil at a Raman shift of 1,655 cm-1 was measured to determine the conversion of the oil. A Gaussian distribution deconvolution algorithm was used to optimize the spectra for quantitative analysis. Refined canola oil and crude corn oil (4.5 wt% FFA) were used to test the robustness of this technique. A high-temperature GC-FID analysis was used to validate the Raman method.

472064 Hydroprocessing of Biocrude into Hydrocarbon Liquid Fuel

Abstract:

Catalytic hydrotreating (HDT) of bio-oil offers a strategic opportunity to produce refinery feedstocks or infrastructure-compatible finished fuels. Nevertheless, HDT of raw bio-oil has proven to be challenging due to the complexity of the reactive oxygenated compounds; the thermal instability of the raw bio-oil cause issues such as reactor plugging and catalyst coking/deactivation. RTI is developing an advanced biofuels technology that integrates catalytic biomass pyrolysis and hydrotreatingÂ to produce hydrocarbon-based biofuels. The goal is to produce a more thermally stable bio-crude with lower oxygen content compared to typical bio-oils. Results from several HDT studies on bio-crude from RTIâ€™s 1 ton per dayÂ catalytic biomass pyrolysis unit show promising results and suggest that the bio-crude oil can be successfully upgraded without a stabilization step.

The main goal of this work is to understand how bio-crude quality affects HDT by hydrotreating several well characterized bio-crude oils. The presentation will discuss in detail, the effect of biocrude composition and process condition (temperature, H_2 pressure, H_2/Oil ratio, and LHSV) on product yield/quality and hydrogen consumption. A detailed knowledge of the impact of these factors can result in innovative technology solutions to control reaction selectivity and minimize hydrogen demand.

462656 Hydrodeoxygenation and Hydroisomerization of Algae Oils to Hydrocarbon Fuels

Abstract:

Microalgae are a promising feedstock for renewable fuels and chemicals because of their potentially high growth rate and lipid content. Recent interest in converting the lipid fraction into completely deoxygenated jet- and diesel-range fuels has motivated research and development efforts in catalyst and process development for deoxygenation and isomerization steps for these lipids, but literature data for algal lipid composition with respect to its effects on DO and HI is sparse. In this work, we report detailed characterization of the extracted lipids as well as deoxygenation (DO) and hydroisomerization (HI) of the lipids extracted from mid-stage (high-carbohydrate content) *Scenedesmus* algae. Over a Pd/C catalyst, nearly complete deoxygenation is achievable, and the DO product is readily isomerized over a Pt/SAPO-11 catalyst. The isomerized product has boiling points primarily in the diesel range and cloud points < -10 °C. Minimal cleanup of the oils is required prior to catalytic upgrading, but bleaching with SiO2 or H3PO4 results in a product with lower nitrogen content.

460978 Continuous Hydrodeoxygenation of Liquid Phase Pyrolysis Oil

Abstract:

Continuous Hydrodeoxygenation of Liquid Phase Pyrolysis Oil

Liquid phase pyrolysis

Non-renewable energy resources such as oil and coal will not suffice the increasing demand of energy in the future. To fill the upcoming gap a biomass liquefaction concept has been developed. This concept consists of two main process steps. In the first step lignocellulosic biomass is converted into pyrolysis oil and pyrolysis char through the bioCRACK process[1]. In the second step the intermediate products are upgraded. Two upgrading technologies were investigated: hydrodeoxygenation (HDO) of pyrolysis oil[2] and hydrogenation of pyrolysis char. The bioCRACK process was successfully operated in pilot scale (100 kg/h lignocellulosic feed) over two years.

Continuous HDO of liquid phase pyrolysis oil

Liquid phase pyrolysis oil was upgraded by continuous hydrodeoxygenation in lab scale. Hydrotreatment was performed catalytically in a plugflow reactor. Operation parameters were 400°C and 121 bar hydrogen pressure. Lab scale results showed a carbon yield of 38% in the fuel phase with a carbon content of 86,15% and a hydrogen content of 13.35%. Fuel boiling range is shown in figure 1. The boiling range is between diesel and gasoline.

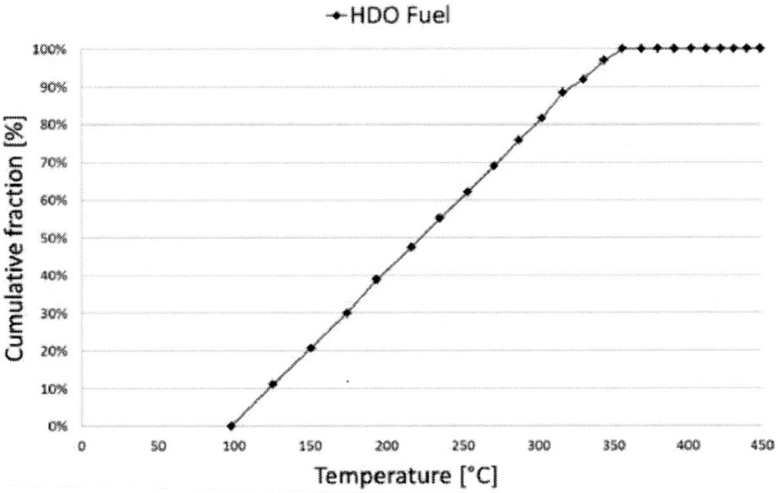

Figure SEQ Figure * ARABIC 1: boiling range of HDO fuel from liquid phase pyrolysis oil

Direct liquefaction of pyrolysis char was carried out in lab scale catalytically and non-catalytically. Non-catalytic biochar liquefaction showed a conversion of 84% and an oil yield of 72%[3], catalytic biochar liquefaction resulted in 100% conversion and 95% oil yield.

The design parameters for pilot scale of the hydrodeoxygenation technology as well as the hydrogenation technology (upgrade of pyrolysis oil as well as biochar) are under evaluation.

ADDIN Mendeley Bibliography CSL_BIBLIOGRAPHY (1) Schwaiger, N.; Elliott, D. C.; Ritzberger, J.; Wang, H.; Pucher, P.; Siebenhofer, M. *Green Chem.* **2015**, *17*, 2487–2494.

(2) Pucher, H.; Schwaiger, N.; Feiner, R.; Ellmaier, L.; Pucher, P.; Chernev, B.; Siebenhofer, M. *Green Chem.* **2015**, *17*, 1291–1298.

(3) Feiner, R.; Schwaiger, N.; Pucher, H. *RSC Adv.* **2013**, *3*, 17898–17903.

AUTHOR INDEX

AUTHOR INDEX

AUTHOR INDEX